LOCKSMITHING

LOCKSMITHING
BY F. A. STEED

TAB BOOKS Inc.
BLUE RIDGE SUMMIT, PA. 17214

FIRST EDITION

FIRST PRINTING

Copyright © 1982 by TAB BOOKS Inc.

Printed in the United States of America

Library of Congress Cataloging in Publication Data

Steed, F. A.
    Locksmithing.

    Includes index.
    1. Locksmithing.   I. Title.
TS520.S73   683'.3   81-18349
ISBN 0-8306-0073-6   AACR2
ISBN 0-8306-1403-6 (pbk.)

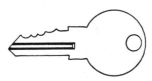

Contents

Introduction

The spiraling rise in burglaries of residences and small businesses, which are normally easy targets, has engendered increasingly open concern. These victims and the general population want to know how to achieve proper security, what proper security will cost and where to find out what is needed to increase the overall security of their residences and offices.

This book traces man's first realization of his need for a means to protect his family and secure his possessions and provides you with the knowledge to improve your security at a reasonable cost.

The many types of locking mechanisms found in the modern home and office and how to install them are thoroughly covered. Ways to identify, repair, and maintain the lock mechanisms; diverse ways you can improve your home or office security; and keying alike and master keying are all covered.

You will learn how to gain entry if locked out and about various techniques experts use to enter a premise without the proper key. Finally, you will learn about the history of safes, how to change combinations, and what details to look for in purchasing a safe. The Glossary of locksmith terms will familiarize you with the terms common to the trade.

With this book you will be able to tackle—with confidence— and successfully complete all the lock work normally encountered in a home and office.

The History of Locksmithing

Recorded somewhere in the history of mankind is the origin of the first lock. The need for a means to protect man and his possessions from unwelcome intruders is as old as mankind. Primitive men made the first effort to meet this need when he rolled a boulder in front of the entrance to his cave dwelling.

Around the sixth century BC, the legend of the lock and key first appeared in recorded history; but long before this, sometime during the beginning of the neolithic age, man learned to cultivate the land and domesticate animals. Approximately eleven thousand years ago, the glaciers began their final retreat, and the earth's climate went through a vast change. During that period, somewhere in the hills to the north of the Tigris-Euphrates valley (where civilization of the western world is thought to have begun) man began his first definite movement out of the caves. During this period in history, man evolved from hunter to farmer. Archaeologists have found farm implements made of flint and pounding stones for milling grain. He learned to fashion tools and possessions too bulky to carry, and to store enough grain to sustain his family between harvests.

With these riches, too valuable to leave unguarded, and with his family to protect, he would block the cave entrance so that no wild animal or raider could enter to harm his mate and offspring or devour his grain.

His lock was a large boulder which he would roll across the entrance to his cave dwelling, all the while checking to insure that he was not observed. He would then quickly plant a wedge-shaped stone beneath the boulder and carefully return the earth to its original appearance.

Upon his return, he would again check to insure he wasn't being observed and would remove the wedge. With his spear or a nearby branch for leverage, he would then roll the stone away. If he feared an intrusion, he would simply roll the stone into position behind him, this time planting the wedge on the inside to prevent an outsider from moving the boulder. It was a primitive, but effective security system.

Locksmithing is one of the oldest of all handicrafts. Locks and keys were a part of folklore early in recorded history. Religion and mythology abound with references to the lock and key. The Babylonian god Marduk is said to have possessed both the keys and the gates to heaven. Throughout, the *Bible* makes reference to keys. In Matthew, Jesus said unto Peter, "I will give unto thee the keys of the kingdom of heaven." The Book of Revelations relates, "The key of the bottomless pit was in the hand of an Angel." The clearest reference, Isaiah XXXII, 22, says, "And the key of the House of David will I lay upon his shoulder; so he shall open, and none shall shut; and he shall shut, and none shall open."

Homer, in 850 BC, wrote in his epics of a crooked key used by Penelope to slam home the bolt to stay her suitors in her husband's absence.

Pliny the Elder, the author of Natural History—a volume of twenty thousand facts, which he presented as a gift to Emperor Titus—relates that in the sixth century BC, Theodous of Samos, an architect and noted sculptor, is said to have created great and ornamented locks and keys for his wealthy patrons.

The first lock of which there is undisputed physical evidence was found during archaeological excavations of the ruins of the great palace of Emperor Sargon II. In 1842, Paul Emile Botta, while French consul at Mosul, uncovered the ruins of the palace known as Sargonsburg, home of the Emperor of Persia, who reigned from 722 to 705 BC. This lavish palace, said to have covered twenty five acres and could house eighty thousand persons. It was located at Ninevah, a suburb of Khrosabad, then the capital of ancient Assyria.

Digging carefully, Botta's crew unearthed a huge ornate hallway at the end of which stood two tall winged statues, each with the body of a bull and the head of a man. Hidden behind the statues was a solid wooden gate closed by a heavy wooden lock.

The wooden lock was mounted on the interior of the gate. It had no keyhole and could not, therefore, be opened like a normal

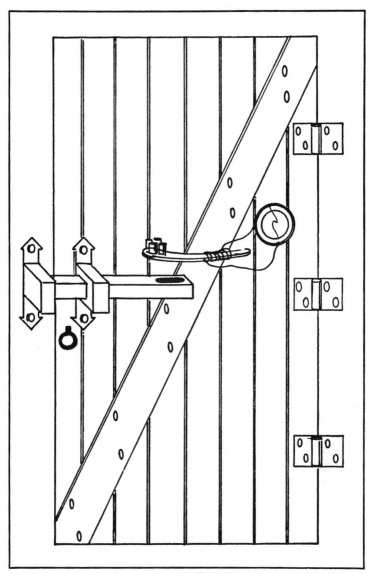

Fig. 1-1. Method used to open the ancient Assyrian wooden lock.

lock. A person approaching the entrance would see only a circular hole about the size of one's arm cut into the gate, normally directly above where the lock had been affixed to the inside (Fig. 1-1). A person authorized to open the lock would put his hand through the hole and insert a large wooden bar—the shape of an outsized toothbrush (Fig. 1-2) and so huge that only an adult could carry it—into a hollow space in the top of the locking bolt. This caused the wooden pins in the lock to lift, freeing the bolt. It would slide back and the gate would open.

Unless one possessed the proper key, the gate could not be opened. Locks using this ancient pin/tumbler principle were fully developed by Yale in 1860 and are still used today in rural Egypt and India. This principle, with refinements, is also used in today's modern pin tumbler locks.

The wooden lock found in the ancient palace by Botta was not invented during Sargon's lifetime. Locks of this design have been found etched by ancient Egyptian artists on the columns of a temple at Karnack constructed around 2000 BC. The lock found in 1842 at Sargonsburg was thus already 1300 years old in its design. Other keys similar in shape and intended for carved wooden locks have been found in the tombs of the dead at Luxor. Many of these keys, made for private homes, are carved from teakwood, with iron or wood pins and ivory handles inlaid with silver and gold. The wooden lock found painted on the columns of that ancient temple about 2000 BC is over 4000 years old and is said to be the oldest known lock in existence.

The Phoenicians, the worlds' first great traders, are believed to have brought the wooden Egyptian lock to the shores of England about three thousand years ago. The Phoenicians came to Cornwall, England, in search of tin, the alloy that converts copper to bronze. The English copied this lock in their native oak. The locks were so strong when made of oak that they remained in use throughout England and northern Scotland until Victorian metalsmiths began to fabricate locks in metal. Wooden locks can still be found in use today on many survival huts in the Alps.

During the days of Homer, the Greeks tied their doors fast with intricately knotted ropes, so cleverly tied that only the authorized owner could find the correct procedure for unknotting them. Superstition during that time prevented all but the most daring thief from tampering with these ropes, lest a curse befall. Greed soon overcame thieves' superstition's. However, as time passed the Greeks, possessing wealth and wishing to protect it, were forced to abandon the rope lock and develop a more substantial locking device. They developed a wooden locking bar and key, a possible variation of the Egyptian lock but much less complicated. Mounted on the inside of the door, it was invisible from the exterior of the building it protected. Unlike the Egyptian lock which required an arm-size hole, the Greek's lock required only a small neat hole for the key slot.

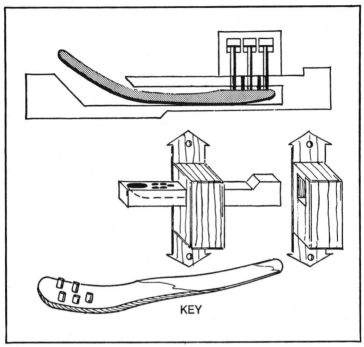

Fig. 1-2. Detail of method used to open the ancient Assyrian wooden lock.

Through the slot was inserted a semicircular iron blade a foot or more in diameter. The blade was equipped with a long handle, similar in appearance to a farmer's sickle. Once the owner, equipped with the proper key, engaged the bolt through the slot, he would then give a half-turn to draw back the bolt. Upon entering, he would close the door, locking it with the key, which was then hung on the wall until needed again.

Such a lock offered little in the way of security. To defeat this lock, an enterprising burglar had only to equip himself with a selection of keys; there was an excellent chance that one would fit the door. Since there were no wooden pins to hold the bolt in position, he had only to get the proper length and curve of the key to pick the lock. The lock was only effective against thieves insofar as individuals poor enough to consider stealing hardly had the price of the iron necessary for the keys.

The Romans fabricated the first metal locks, using iron for the lock itself and bronze for keys. They would fasten a metal plate with a keyhole to the front of a door; usually the door was made of wood. The locking mechanism was then affixed into channels cut into the

5

door frame behind the plate. The Romans invented the wards inside the lock which prevented all but the properly-cut key from entering the lock. Realizing the importance of making different locking mechanisms that would not work by a simple key, they formed springs to press the pins into the bolt. They also used different sized pins for centuries, wards were the primary security of locks. Few Roman locks have ever been found. Constructed of iron, they rusted hopelessly into formless masses, or corroded into usually undistinguishable lumps.

Roman keys are an entirely different story. Antique Roman keys are found in important collections throughout the world, collected by nobility and the common folk. The mystique of the key, which was mentioned throughout the Bible and in mythology, created a desire for the key's possession.

Because Roman garments had no pockets, the Romans developed small keys that could be worn on the fingers. These key rings were tiny and delicate and may have been worn by women. Most large keys used by the Romans were carried loose or on a circular device; it was usually entrusted to servants.

The principle of warded locks was eventually applied to portable locks (later called padlocks). Although the Chinese developed the padlock, the Romans are credited with the development of the ward principle used within the padlock. The Romans also popularized the padlock throughout the known world.

Padlocks were often quite clever in their design. Early Chinese and East Indian cultures used ornamented metal padlocks designed as idols, flowers or animals. The keyhole was often cleverly concealed for added security. As civilization spread across Europe, padlocks were devised that ejected needles, poison darts or knives if a foreign key was inserted.

The ancient city of Pompeii was engulfed by an eruption of the volcano Vesuvius in 64 AD. Centuries later, archaeologists unearthing the buried city came across a building that was the site of a locksmith shop. Here they found many types of door locks, padlocks and highly decorated keys, some inlaid with gold and silver. Oblong instruments and various other tools were also discovered; possibly these were picks used by the ancient locksmith to open his patron's doors in an emergency.

After the decline of the Roman Empire, the world entered into the period known as the Middle Ages. In the thousand-year span that followed, few changes were made in lock design. The simple warded lock remained the basic mechanism. For added security, locksmiths turned to elaborate wards and intricate designs. Locks became works of artistic creation. These were the days of knights and fabulous castles full of treasure. It was a point of pride and prestigue to have the most elaborate and ornate locks guarding the castles.

The first locks constructed completely of metal began to appear during the reign of Alfred the Great, who was a collector of fine locks. Exactly when all European locksmiths began making locks entirely of metal is unknown, but it was probably not much after 900, when the reign of Alfred the Great ended.

The beginning of the Gothic period (middle of the twelfth to the early fifteenth century) finds the craft of locksmithing considerably advanced from the days of the Romans. While Gothic keys continued to be constructed of bronze and iron, they were, at this time, mostly flat, with elaborate bows forged in the shape of diamonds. During the early Gothic period there was little noticeable advancement in the locking mechanism. The first door locks of this period had horizontal rather than vertical keyholes. The horizontal keyhole kept the mechanism simpler—no extra lever was necessary to actuate and retract the bolt. In the latter part of this period the keyhole was changed to the present-day vertical keyhole. This change called for a more complicated internal locking mechanism, but Gothic locksmiths proved themselves up to the task.

German, Italian and French craftsman of this period produced particularly excellent locks. Germany esteemed the lock craft so highly that Charles IV established the title of Master Locksmith.

During the thirteenth and fourteenth centuries, Gothic locks offered fine forged ironwork with designs of animals and flowers on the lock cases. In the early Gothic period, the craftsman melted iron, cut ingots into bars, and then pounded the bars into sheets. He then worked the metal until he had produced rough shapes. He forged lock bodies and added ornamentation, engraved or etched until the end product stood as beautiful testimony to his craft.

Among the fine lock craftsman of this period were men like Jorg Heusz (a master clockmaker) and Hans Ehemann (reputed to be the inventor of the letter-ring combination lock), both of Nuremberg. These men, along with others equally skilled, were invited to make locks and keys for French and Italian royalty. The moving parts of the locks these men developed were closely fitted and finished, and the lock cases were lavishly decorated. Even the keys were works of art! But the security within the lock was still dependent upon elaborate warding.

Until the Renaissance, there were no springs in the locks which would slam the bolt in place when the door was closed. Until then, locks required that the key be turned to lock as well as unlock the door. The self-locking bolt, which developed during the early Renaissance, had a beveled end and a spring. When the door was closed, the bolt would snap into place.

Locksmiths began to spend less time forging, and more time working with the file, hammer, cold chisel and saw. Freed from his dependence upon the forge, the craftsman could proceed slowly, concentrating on details. Lock cases lost their rough hammered look, and became extensively decorated. Key bits remained relatively the same, but the bow became much more a work of art. Keys were carved with knights slaying dragons on the bow. The end of the Gothic period and the beginning of the Renaissance saw the rise of the Locksmiths Guild. Before a journeyman locksmith could be accepted by the Guild of Locksmiths as a master of his craft, he was required to submit for approval at least one individually fashioned lock and key. These demonstrated his skill with the chisel, saw and file. The locks and keys were then placed on display in the Guild Hall.

THE CUSTOMS AND TRADITIONS OF LOCKSMITHING

The great medieval guilds of locksmiths, which had their formation during the middle ages, were all-powerful. The guilds regulated the terms of apprenticeship as well as devised the rules and conduct of its journeymen. They required apprentices and journeymen to submit the carved lock and key for approval before accepting them as master locksmiths.

The guild completely controlled the techniques of the masters, regulated prices, and determined the construction of locks down to the number of rivets used. The guild was supreme master over the craft; the penalty for defiance was expulsion from the guild and loss of the right to practice one's trade.

At first, the strictness of the guildmasters helped maintain the integrity of the locksmith's trade, but, like all autocratic organizations, the guild became arrogant, jealous of its power over its members, and unwilling or unable to advance.

As time went on, locksmithing became a father-to-son enterprise, following guild guidelines as late as the nineteenth century. Restraints placed on the master by the guild meant that locksmithing did not progress beyond the addition of intricate designs to confuse would-be thieves.

False keyholes, phony wards, and gingerbread design were added to the lockbodies. Money chests that cut off the thief's fingers and locks that fired bullets or ejected knives when tampered with were part of the locksmith's stock in trade. Secret panels and hidden keyholes were used to foil the burglar. Throughout this period in history the basic construction of the lock mechanism remained the same as in the days of the Romans.

LOCKS AND KEYS IN EUROPE

German craftsmen devised an extremely complicated lock for chests which used many different bolts which, when locked, would shoot out in every direction. Multiple keyholes and several locks were fairly common. One money chest was equipped with twelve bolts operated by several keys. Some locks were armed with steel barbs; if anyone but the authorized owner, who knew the secret combination, attempted to open the lock, the steel barbs were ejected by a strong spring, piercing the intruder's hand.

This trick had numerous variants. For instance, many of the massive money chests, in common use in castles, churches and monasteries, were equipped with an inner lid beneath the chest lid. The thief skillful enough to master the intricacies of the locks or strong enough to force open the top of the chest would find beneath the locked top a second lid resembling a close-fitting tray. This lid would be equipped with finger holes to lift it out. What he did not know was that the underside of this innocent looking tray consisted of a mechanism resembling a small animal trap. When he inserted his fingers into the holes, he would trigger a pair of saw-tooth jaws which would lock his fingers beneath the tray. The pain of the trap was usually sufficient to cause the thief to cry out, thereby alerting the guards or servants. But even if his cry did not arouse anyone, he was trapped: he had no alternative short of cutting off his own fingers.

The only notable improvement added to locks during the Renaissance was the self-locking bolt and the single lever tumbler (which would hold the bolt in place so it would not move forward or backward unless the proper key lifted the lever). These sound principles were never fully appreciated by Renaissance craftsmen who handed it down to their more skillful heirs of the eighteenth century.

During the eighteenth century, English locks came into their own, usually consisting of a steel key and locking mechanism and a sculptured brass case.

Around the beginning of the Industrial Revolution, locksmiths turned their attention to developing the inner works (the locking mechanism). Until then, more effort had been spent on decoration and concealment of the keyhole than improvement of the lock itself.

In 1778, Robert Barron of England patented a lock which no longer depended entirely on the wards around the keyhole (although it did retain them). He designed two spring-loaded lever tumblers of different sizes to check the bolt. To operate this lock, one needed a key forged which would not only get past the wards but which, having proper cuts in the key bit to lift levers to the proper level, would release the locking bolt. This multi-lever tumbler lock was, from a security standpoint, a vast improvement over its predecessors. The key looked simple and the cuts resembled those of earlier keys. But, with the wards and lever tumblers, the lock was relatively complicated. A new era in modern lock construction had now emerged. Barron's lock, however, had one weakness: the lower edge of the two levers were, instead of being uniform height, at different levels. A locksmith named Joseph Bramah soon discovered this and found a way to pick the lock.

Using what is commonly known as a wax-impression, he would first select from his stock a key capable of entering Barron's keyway. He would then spread a thin layer of color-wax along the edge of the bit parallel to the stem. Then, exercising extreme care, he inserted the key and turned it until he felt the first hint of resistance. Withdrawing the key, he filed off a fraction of the metal below the point marked in the wax when the key made contact with the lower edge of the lower lever.

Again and again the locksmith would patiently apply wax, test and file where marked (removing no more than a fraction each time); he soon found the resistance he experienced was from contact with the lower edge of the higher lever. He repeated the procedure on the second marking—filing the bit a fraction each time—until at length he found the impression on his line of wax to be equal along the edge. He had thus determined the actual levels of the lower edges of both levers. Removing only a small amount from the key bit he soon had an exact duplicate of the original key. This locksmith's procedure is commonly known as "keying by impression." He then inserted the key; when it turned, it engaged the bolt and the cuts engaging the levers; he continued to turn the key, retracting the bolt, and successfully picked the lock.

Joseph Bramah's name is one of the most prominent in the entire history of locksmithing. In 1784 he patented a lock which he called the Bramah Lock. Being somewhat of a showman, he advertised his lock as being as "impregnable as the Rock of Gibraltar" and offered a substantial reward to anyone who could pick it.

The Bramah key was a small metal tube with narrow longitudinal slots cut in the end. When the key was inserted in the cylindrical keyway of the lock, it depressed a number of metal slides to the depth controlled by the slots. Only when all the slides were depressed to the exact right distance would the key turn and the bolt retract. The Bramah lock was the first high security lock to be key-operated and Bramah's concept, with refinements, can be found in today's Ace cylinder lock. Other notable locksmiths followed the lead of Barron and Bramah. Locksmithing began a new page in its long history. Names such as Chubb (three brothers who used the lever tumbler principle to patent many improved locks), Andrews, Pettit and Parons are recorded in connection with improved design and security.

The Bramah lock went unchallenged until 1851 when, at the International Industrial Exhibition in London, an American locksmith named Alfred C. Hobbs, using picks of his own design, succeeded in opening both the Bramah and the Chubb locks in less than half an hour.

Before Hobbs was through that day he also picked a newer and larger lock invented by Bramah as well as a Cotterill detector lock. The detector lock had a weighted part which was mounted so that it would tip into a locked position if the wrong key was inserted or if someone tried to pick the lock. Once the weight had been shifted by an unauthorized attempt to open the lock, only the correct key could open the lock. Hobbs later made improvements on the Cotterill lock.

Hobbs found England so much to his liking that he established his own lock and safe company there and stayed on for thirty years. In 1854, one of Chubb's locksmiths succeeded in picking one of Hobb's patented Protector locks. The locksmith had studied the Protector lock for months and had constructed special tools to pick it.

LOCKS AND KEYS IN THE UNITED STATES

The first settlers to arrive on the eastern shores of America had no locks. Their first homes were caves or simple sod houses, temporary shelters until they could cut timber and raise log houses.

As they began to prosper, they built new and better dwellings, many furnished with goods imported from England. But lock development trailed behind the technical level prevailing in Europe. The better locks were imported. Doors on most homes had only latch string locks with the strings hanging outside. A visitor would pull the string to open the door (from which comes the saying "Our latch is out for you"). At night the homeowner simply pulled the string back inside, fastened it securely and thus prevented entry.

As industry in the United States progressed, locksmiths copied the English locks and, in some cases, improved upon them. In 1836, Solomon Andrews, a locksmith in Perth Amboy, New Jersey, developed a lock with both adjustable tumblers and keys, allowing the owner to change the positions of the tumblers inside the lock at will. The owner, after making his tumbler adjustment, would then adjust the moveable bits on his key in order to work his lock. In theory, he could do this as often as he wished. But in practice, it required dexterity, patience and a minimal knowledge of the lock's inner workings. Andrews had developed his changeable lock not for the homeowner, but for banks and businesses. An important advantage was that, if a key were lost or stolen, the lock's owner could change the lock and key so that only his key would match the new combination.

The demand grew for more and better locks with which to secure strongboxes and protect homes and businesses. Locksmiths responded to the demand. J. Perkins of Newbury, Massachusetts invented a complicated combination bank lock with a key of movable washers which could be rearranged at the owner's fancy. Day and Newell of New York developed the Parautoptic Bank Lock (from the Greek, meaning "hidden from sight"), which consisted of two complete sets of changeable tumblers. This lock concealed its inner workings so a would-be thief could not examine it; if the thief succeeded in getting into the lock mechanism, a detector plate would fall down, thus closing the keyway. In 1847, Newell's Parautoptic lock won a gold medal for mechanical excellence at the National Mechanic's Institute of Lower Austria. This was quite an honor for an American invention.

In 1848, the Herring Company developed a "grasshopper lock" for safes. The key for this lock consisted of a semicircular piece of metal equipped with teeth. To open the lock, one inserted the key into a slot at the left of the knob and turned the handle. As the lock released, the key was ejected from the lock; hence the name "grasshopper."

Linus Yale, Sr., a successful bank lockmaker, was, of course, not the first to construct a lock using the ancient Egyptian falling pin principle. But his 1844 invention of the pin tumbler cylinder lock combined the Egyptian falling pin concept with the Bramah cylinder design, thus creating the most secure lock to date. Alfred Hobbs described in 1853 how " .... the lock is similar to the Egyptian with something like the Bramah." Hobbs, while describing the lock's inner workings, succeeded in picking it; but he apparently admired it nonetheless. It had two cylinders, one contained within the other. It was held together by a series of pins and springs protruding through the cylinders into the keyhole, which was in the center of the inner cylinder.

The elder Yale constructed his lock with four sets of pin tumblers, but Hobbs noted that the lock could be produced with anywhere from one to forty pins. It would present an insurmountable barrier to anyone but a thief highly skilled in picking.

Linus Yale, Jr., chose as his first career to be an artist, but soon turned to inventing and naming locks. During his lifetime he developed and patented locks such as the Yale Infallible Bank lock in 1852 (with a changeable key). He invented the Yale Magic Bank lock and the Yale Double Dial Bank lock set, the standard for present day combination locks. Each dial controlled a separate set of four circular tumblers; each tumbler had one hundred possible changes, with a total number of possible combinations well over one million. Yale, Jr., spent years improving the design of the pin tumbler cylinder lock so it could be mass produced and still provide excellent security for everyday uses. The Yale pin tumbler design has won worldwide approval, as it is a very secure lock and reasonably inexpensive to manufacture. It is also easy to install and fairly easy to rekey. Its keys are a flat-and-grooved type of cylinder key which today stills bears Yale's name.

Today most pin tumbler locks are manufactured with anywhere from three to seven pins. Pin tumblers are used today wherever a moderately high level of security is desired, but not the highest level possible. Lever tumblers are used for bank deposit boxes, in mental institutions, and in some prisons. They are used because the key blanks for lever tumblers are limited in quantity; some types are controlled by the lock manufacturer. Pin tumbler blanks, on the other hand, are readily available from locksmiths, hardware stores and even some grocery stores.

James Sargent, who developed the magnetic lock in 1866 and perfected the first model of his time lock in 1873, was another American lockmaker of stature. The concept of a time lock had been dreamed of ever since the days of the Medieval locksmith guild, but

Sargent was the first to develop a practical model. In the mid 1800s, thieves began to kidnap bank officials and force them to open the vaults. Sargent's time lock quickly put an end to this practice.

As modern mass production techniques were introduced, the locksmithing craft divided into two distinct groups: lock manufacturers and lock repairmen. The manufacturing end of the business became highly technical and complex, eventually leading to the creation of a multi-million dollar industry. The term "locksmith" now refers to the lock repairman rather than the manufacturer.

LOCKS AND KEYS IN THE TWENTIETH CENTURY

In the twentieth century several new inventions have increased the scope and efficiency of locksmiths. One of these, invented by Henry Gussman in 1909, is the key cutting machine for duplicating keys. This machine ended the tiresome task of hand-filing keys, and it made exact duplicates of original keys in a matter of minutes. In 1926, engineers at the Independent Lock Company developed the key code cutting machine. This machine allowed locksmiths to cut keys for locks using nothing but the lock serial numbers; these numbers determined the depth of each cut.

There were other inventions by such men as Hoffman, Johnston and Epstein. Their tools, devices and methods enable locksmiths to measure lock tumblers without disassembling the lock and provide an effective means to open locks in emergencies. Another famous name is Walter Schalge, who in 1920 pioneered the first pushbutton locking mechanism within the door knob, and also perfected his own design of a cylindrical (key-in-knob) locking mechanism.

With today's splendid engineering techniques, which make close tolerances possible in machining locks; with the development of locks which respond to electronic signals; and with the invention of locks which utilize magnet-imbedded plastic cards to trip electrical circuits, the ancient Egyptian falling pin tumbler concept still remains the primary locking device. Today it contains such modern refinements as spool (sometimes called mushroom) pins, and the latest interlocking pin tumbler using a key cut and controlled by the manufacturer, but the essential concept remains the same.

Equipping your home or office with the proper security protection is much simpler and less costly if you remember that you can only hope to *lower*—not eliminate—your chances of victimization. The bulk of responsibility for your protection is yours alone.

On a national average, most burglaries are committed by kids. Against them, normal precautions—such as good solid doors, adequate locks, properly-fitting frames, and locks instead of latches on windows—are sufficient.

Most Americans spend needless millions on inadequate security. Proper protection need not be expensive. After you determine the inadequacies of your doors and windows, you can either pay a locksmith to correct the deficiencies or, with a little patience, you can correct them yourself.

Types of Locks

Since the beginning of time, man has devised methods to guard his possessions and protect his family. The rock used by primitive man to close the entrance to his cave dwelling, wooden locks that secured the ancient gates of Egypt, locks of iron and brass introduced by the Romans, and the latch string locks used by our colonial ancestors are all examples.

Today, most homes contain a variety of lock types. Warded-bit locks are used on interior doors and storage shed doors. Lever locks are found on antique desks, commodes and storage chests, high-quality desks, and, of course, safe deposit boxes. The disc tumbler lock may be found on almost all automobiles as door locks, ignition switch locks, and glove/trunk locks and on office file cabinets, vending machines, and doors. The pin tumbler lock, the most common lock today, can be found in the doors of most modern homes, office buildings, hotels and motels, public buildings, and on desks and automobiles.

THE WARDED LOCK

Warded locks were invented by the Romans. The wards inside the lock prevented all but the proper cut key from being rotated inside the lock. For centuries, wards were the primary security of locks.

Throughout the middle ages, skilled craftsmen, especially the German metal craftsmen of Nuremberg, were employed in the making of metal warded locks. The moving parts of the locks were closely fitted. The exterior casings were lavishly ornamented, and even the keys were works of art. Lock security, however, was still dependent upon warding. Sometime during the Renaissance a lever tumbler was added to the warded lock to improve security (Fig. 2-1). The key not only passed the fixed obstruction, but it also lifted the lever to the proper height necessary to retract the bolt. Nevertheless, with a little practice, a skilled thief could soon manipulate the lever and pick the lock.

The *Warded-bit key lock* (the full name of the device) is the simplest in construction of all locks in use today and can still be found in old homes and apartments. The device as we know it, is over one hundred years old. Originally used as a primary, exterior door lock, it is today used on interior doors such as bathrooms and bedrooms.

Warded locks derive their name from the word *ward*, which is defined as the action or process of guarding. Simply put, they consist of a projecting ridge of metal in a lock casing or keyhole permitting only the insertion of a key with a corresponding notch. The number of wards per lock may vary depending on the quality of the lock. Two is the minimum, although the cheaper warded padlocks may have only one ward.

Constructed of cast iron casing and lock assembly, the lock usually consists of two main parts. One part holds the locking mechanism; the other is the cover plate (Fig. 2-2).

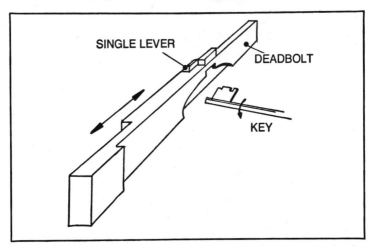

Fig. 2-1. The warded lock as it appeared during the Renaissance, after the addition of the lever tumbler.

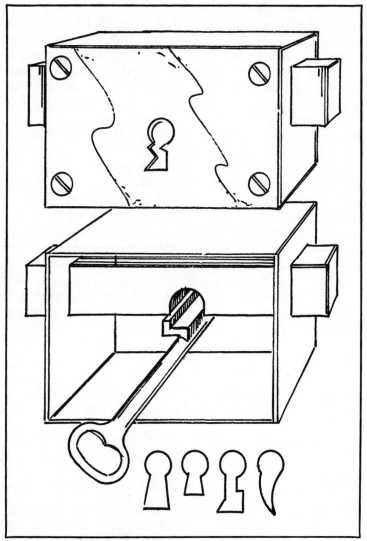

Fig. 2-2. Types of warded locks.

There are two types of warded locks, and they are readily identifiable by their installation (Fig. 2-3). The *rim lock* is installed on the surface of the door. The *mortise lock* is installed in a hollow space in the edge of the door. Of the two, the mortise style is the most desirable because it is neater in appearance.

Generally, only the rim lock is equipped with a safety latch (Fig. 2-4). This locks the spring lever latch, preventing entry from

18

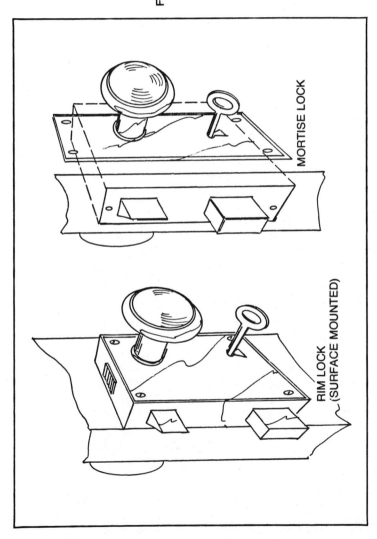

Fig. 2-3. Rim and mortise locks.

MORTISE LOCK

RIM LOCK
(SURFACE MOUNTED)

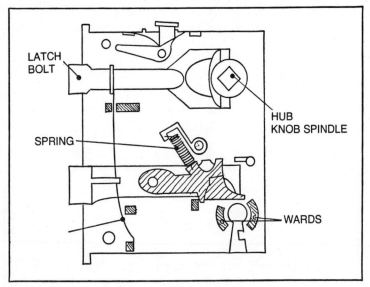

Fig. 2-4. Rim lock equipped with safety latch.

the outside without the proper key, but allows the door to be opened from the inside by a simple turn of the door knob.

Some of the obvious weaknesses of the warded bit lock are the relative ease with which the wards or obstructions can be overcome and the limited number of keying possibilities.

The security of the lock was in its cast iron construction which, if properly maintained, would last forever, needing few, if any, repairs.

Key changes were limited to approximately fifty different keying combinations. Thus, there was a good chance that one in fifty persons possessed a key that would fit a particular lock. Actually, because of wear and tear on both lock and keys, this number was even further reduced. Warded bit locks can be picked in a matter of seconds by an expert using nothing more than a bent section of a clothes hanger or a nail. A person with little knowledge of the lock's construction can pick a lock of this type in less than an hour. When found in use today, these locks are usually backed up by a rim deadlock or some other locking device, or are in use as interior door locks.

Operation

The operating principle of the warded lock is simple (Fig. 2-5). First, one views a typical keyhole and a key fitted for that particular

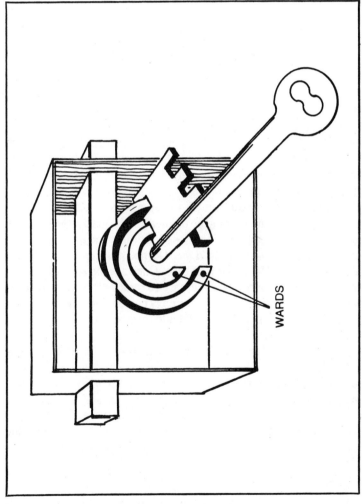

Fig. 2-5. Operating principle of the warded lock.

WARDS

21

lock. Note the metal obstruction jutting into the keyhole which allows entry of only the properly-cut key. Next the key bitting cuts (Fig. 2-6) allow the key to be turned in a circular motion, engaging the dead bolt and drawing it into the lock casing. To lock the door the procedure is reversed.

Maintenance and Repair

Removing Paint. One of the primary causes of lock failure in the warded lock is ordinary house paint. The homeowner fails to take the time to paint around the lock. He paints over it and, when the paint dries, the exposed parts become rigid. To prevent this, tape around the exposed lock parts before painting doors and trim.

To repair such a lock, first remove the lock from the door, being careful not to crack the new paint. Then scrape the paint off with a paint scraper or knife blade. *Do not use paint remover.* Although paint remover will soften paint, it dries quickly and leaves a residue harmful to locks which will in time cause rust.

Lubrication. Never use oil to lubricate a lock. Oil collects dust and grime, a harmful combination to locks. Instead, use graphite, preferably in powdered form. Graphite is a dry lubricant and is preferred by most professional locksmiths. Spray it lightly into the lock. The fine dust will settle on the lock parts and cling to them. Use it sparingly and be careful not to get it on your clothing as it is very difficult to wash out.

Regular Maintenance. To keep a warded bit lock in peak operating condition, it should be opened yearly and dusted out, then given a touch of graphite and reclosed. At the same time, check the strike plate to ensure it is tight. Do not overtighten the wood screws when replacing the lock case. Once a wood screw has been overtightened, it becomes necessary to remove the screw and dab some plastic wood compound or expoxy cement into the hole with a toothpick, using just enough to rebuild the wall, but not so much as to close the screw hole.

Obtaining Parts. Parts for repair of some warded bit locks can still be found in hardware stores. Parts can also be located in junk shops, antique stores and in wrecked or condemned houses. When looking for parts for a warded bit lock, be sure to carry your lock with you and fit the parts right then and there. Over the life span of the warded bit lock, many manufacturers have come and gone and, in the process of development and manufacture, many changes have been made.

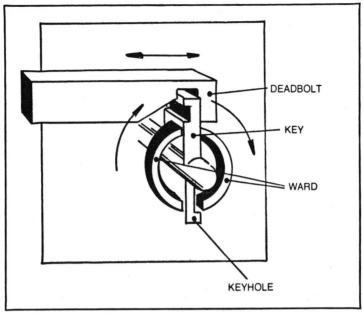

Fig. 2-6. Detail of warded bit lock.

THE LEVER LOCK

Sometime during the fifteenth century an unidentified locksmith developed the first *lever tumbler lock*. This development constituted an improvement over the warded lock, but a very small one. A thief might find the lever lock insurmountable the first time, but after two or three attempts it would present little more problem than the warded lock.

The first lever tumbler lock had only one lever. To operate this lock, the properly cut key would lift the lever to a certain height, turning the key would then retract the deadbolt (Fig. 2-7). The single lever concept was popular almost to the present day, and locks based on this concept can still be found in many older homes. Lock manufacturers combined the principles of the single lever lock with the warded lock to increase security.

In 1778, an Englishman named Robert Barron developed and patented a lever lock which improved considerably upon its predecessors in two important aspects. It consisted of more than one lever tumbler, the levers of which operated independently. Each lever had to be raised to a certain point. If lifted too far or too little, the key would not operate the lock.

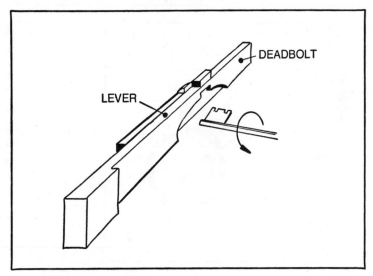

Fig. 2-7. First lever tumbler lock.

By using more than one lever tumbler, Barron was able to construct a lock that was fundamentally sound. For the first time since the ancient Egyptian locks, locking action was applied to the lockbolt. Gone forever was the need for trickery and false designs. A new era in lock construction had begun.

The multiple lever tumbler lock offered a tremendous increase in security. A lock such as this was equipped with ten levers (lever locks were normally built with anywhere from two to ten levers) and manufactured to permit ten different bitting heights. This meant that the lock had a keying possibility from zero to ten billion. In actual practice the number is far less, somewhere on the order of thirty to fifty thousand, but still many times more than the warded bit lock and the single lever lock then so widely in use.

The basic design developed by Barron is still in use today. However, today's lever locks have the levers mounted on the deadbolt. Subsequent lever tumbler inventions, such as Bramah, Newell, Chubb, Andrews, Pettit and Parons, improved on the original design by adding components, but none altered the original Barron invention. These men were inventors who enthusiastically challenged the world and each other to pick their locks. This intense rivalry led, of course, to many improvements in locks. At the London Exhibition in 1851, Alfred Hobbs succeeded in picking open the Barron lever tumbler lock (as well as other locks), considered the best England had produced.

Locksmithing had its heyday in the mid-nineteenth century. Newell's improvements on the lever lock—called a parautoptic lock—consisted of two sets of lever tumblers. The first worked on the second, and included a plate that turned with the key, preventing inspection of the lock's interior. There is no known record of this lock being picked.

Today, lever tumbler locks can be found in the doors of some homes and in such institutions as hospitals and prisons, in attache cases, suitcases, cabinets, desks, fine antique furniture such as rolltop desks (Fig. 2-8), and on high-security safe deposit boxes, normally equipped with fifteen or more lever tumblers and requiring two keys to permit opening.

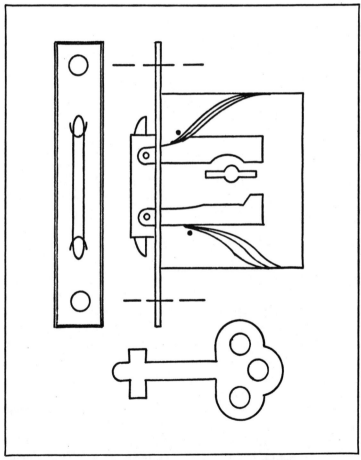

Fig. 2-8. Type of lever tumbler lock found in antique rolltop desk.

THE DISC OR WAFER TUMBLER LOCK

The *disc tumbler lock*, also called the *wafer tumbler lock* by some manufacturers, was developed because of a demand for a small, low-priced lock offering a reasonable degree of security. The disc tumbler lock resembles the pin tumbler lock in outward appearance, it uses a much simpler mechanism and costs less to manufacture.

Disc tumbler locks are easy to install and remove and are commonly used as less expensive substitutes for pin tumbler locks, where security is not as crucial. You may find the disc tumbler lock mechanism in many of today's automobiles, less expensive door locks, metal desk locks, filing cabinets, vending machines, padlocks, cam locks, and locking handles (Fig. 2-9).

Disc tumbler lock keyways usually have simple side ward indentations. The key, which looks similar to the pin tumbler lock key, is usually shorter in length but equally as broad. The disc tumbler key may be cut on one or both sides. The very best disc tumbler lock is less secure than the pin tumbler lock because greater tolerances are permitted in manufacture.

Disc tumbler locks consist of four major parts. These are: plug housing, with disc/wafers and springs; shell; cam or locking bolt; and retainer (Fig. 2-10).

A disc lock may have one disc or eleven discs, but no more than five possible disc tumbler depths are used by most manufacturers; the pin tumbler, by comparison, uses up to ten depths. There is thus less flexibility in keying the disc tumbler. Disc locks are capable of between three and four thousand keying combinations.

Operation

Each disc tumbler lock has a series of chambers which house the flat metal movable discs and springs. For ease of identification, these chambers are given a spacing number which starts at the keyway. These numbers are standard throughout the lock industry. The chamber nearest the keyway is the first spacing and is given the number one. The second chamber is called the second spacing, and so forth (Fig. 2-11).

Spring pressure against the spring wing on the disc forces the disc down so that the lower portion of the disc extends into the locking groove of the shell (Fig. 2-12). The disc then locks the plug and shell together, preventing the turning of the lock.

To operate the lock, the properly cut key, when inserted, must lift the disc out of the lower locking groove. It must be moved up

26

DESK LOCK

LOCKING HANDLE

PADLOCK

Fig. 2-9. Types of disc/wafer locks.

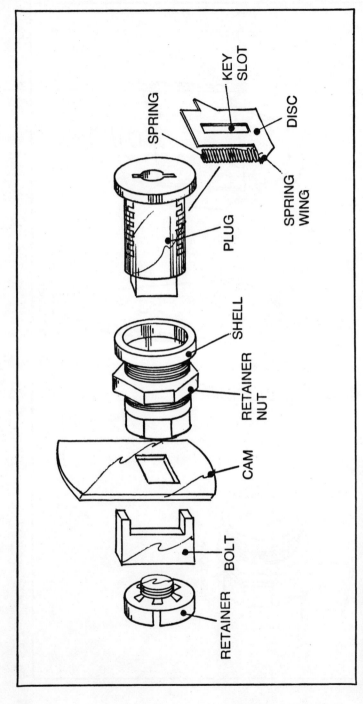

Fig. 2-10. Exploded view of disc tumbler lock.

28

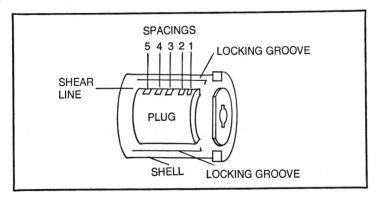

Fig. 2-11. Disc tumbler spacings.

flush with the shear line (i.e., the circumference of the plug, that point where the plug will turn freely in the shell). This allows the plug to be turned. When it turns it retracts the cam or bolt.

THE PIN TUMBLER LOCK

The first lock utilizing the falling pin tumbler concept originated in the Middle East. It is one of the oldest locks known to man and was found in the ruins of the palace of Sargon II, at Khorsabad, near the biblical city of Ninevah. Known as the Egyptian lock, it is estimated to be some four thousand years old (Fig. 2-13).

The lock consisted of three primary parts and was constructed entirely of wood. It used a large wooden bolt to secure the door, one end of which was hollowed out to permit the entrance of a toothbrush-shaped wooden bar with wooden or iron pegs instead of bristles. A hollow wooden assembly attached to the interior of the

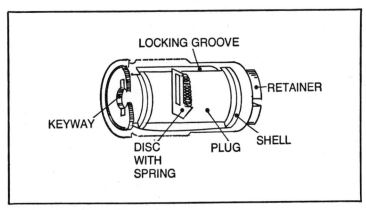

Fig. 2-12. How disc tumbler works.

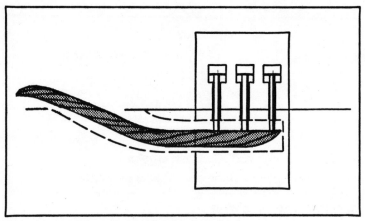

Fig. 2-13. Cutaway view of the Egyptian lock.

door contained compartments which housed several wooden pins positioned to drop into holes perforated into the upper surface of the bolt. The bolt would thus be secured firmly into place. The wooden key (shaped like a toothbrush), when inserted, would align the pegs with the perforated holes; it then simply lifted the wooden pins clear, allowing the bolt with the key to slide back. Locks of this type have been found in Japan, Norway and England and are still in use today in rural Egypt and parts of India.

The falling pin concept was used throughout history. It was developed to its present state in 1860, by Linus Yale, Sr., an American blacksmith turned locksmith. Yale's lock had two primary parts: the plug, which contained the lower pins and received the key; and the shell or cylinder, which housed the upper pins and springs. This concept, with refinements, is still in use today in the construction of pin tumbler locks.

Linus Yale, Jr., an American portrait artist turned locksmith, is the father of the modern pin tumbler lock. He improved upon his father's design, patenting his first lock in 1861 and an improved model in 1865. Yale used a rotating plug slotted to accept pins of different lengths (Fig. 2-14). The edge of the key was cut to the depths corresponding to the length of the pins (Fig. 2-15). When inserted in the keyway, the cuts in the key raised the pins to the same height, freeing the plug to turn in the cylinder and retract the locking bolt.

The lock body used by Yale was not a new invention. The separate pin tumbler cylinder allowed the lock body to be as large as necessary, but the size of the pin tumbler cylinder could be kept

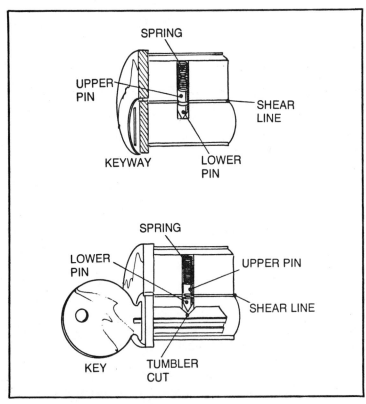

Fig. 2-14. The original Yale lock.

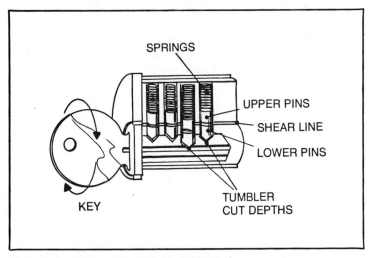

Fig. 2-15. Interaction of key and pins in Yale lock.

31

small and uniform. This concept modernized locksmithing and ended the era of large and cumbersome keys.

By separating the two functions, the pin tumbler cylinder could be mass produced in one operation and the lock body could be produced in whatever size, shape or design necessary for security. These mass production techniques, started by Yale over a century ago, are still used in the manufacture of pin tumbler locks today. Using one assembly line to manufacture the lock body and another to manufacture the pin tumbler cylinders, the two components are then married and packaged for delivery. Each pin tumbler resembles the other in appearance, but is different with regard to the keys operating it.

As time passed, lock manufacturers began to realize how versatile the pin tumbler lock was. Today, the pin tumbler concept is incorporated into numerous lock bodies (Fig. 2-16).

The pin tumbler lock is today's most popular locking mechanism. It is found on the doors to most modern homes, offices, commercial buildings, and on better desk locks, hospital drug cabinets—wherever a lock is needed to provide something less than maximum security. Most locksmiths today agree that no other type of key-operated lock can equal the pin tumbler, relative to cost, size, service and keying possibilities.

The actual service life of the pin tumbler lock is not known, but this writer personally knows of several models of pin tumbler door locks that were still functioning after twenty years of service.

As manufacturing techniques improved, pin tumbler locks were produced with closer tolerances. The result was improved security. Steel bolts, segratted pins, mushroom pins and numerous other invocations were introduced to minimize the possibility of picking, shimming, or physical assault.

Many new locks have been developed in this century, such as locks and keys implanted with bits of magnet (Fig. 2-17). Inserting the proper key would align the metal pins, permitting the plug to turn and retract the bolt. *Spool pins* (sometimes referred to as *mushroom pins*) and interlocking pins were developed by the Corbin Lock Company (Fig. 2-18). These pin tumbler cylinders offer improved security against picking and physical assault. They are often equipped with drill-resistant shields and rods to prevent the would-be burglar from drilling the pins. Some models require that the manufacturer cut the keys.

Numerous modern keyless locks are on the market which respond to electronic logic signals and use a digital counter as a key.

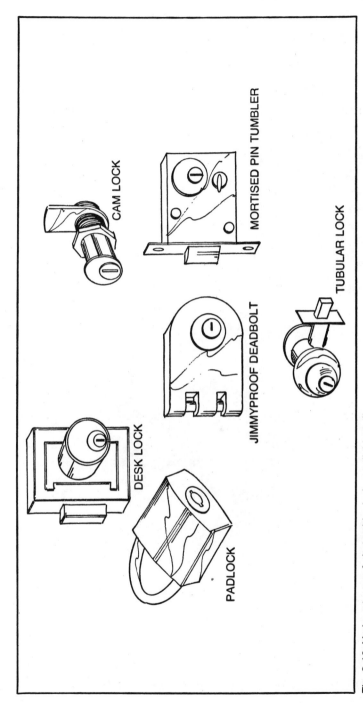

CAM LOCK

MORTISED PIN TUMBLER

TUBULAR LOCK

JIMMYPROOF DEADBOLT

DESK LOCK

PADLOCK

Fig. 2-16. Various types of pin tumbler locks available today.

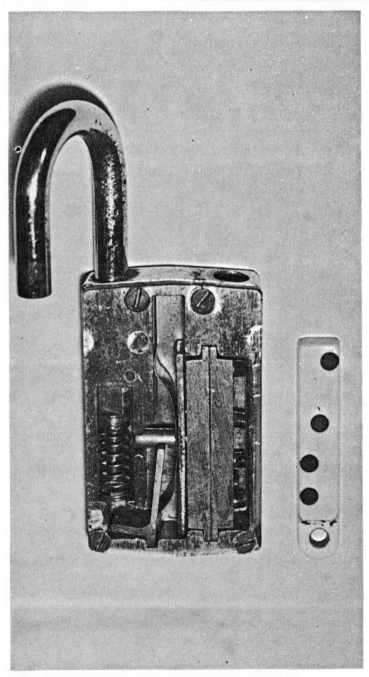

Fig. 2-17. Photograph of a lock implanted with bits of magnet.

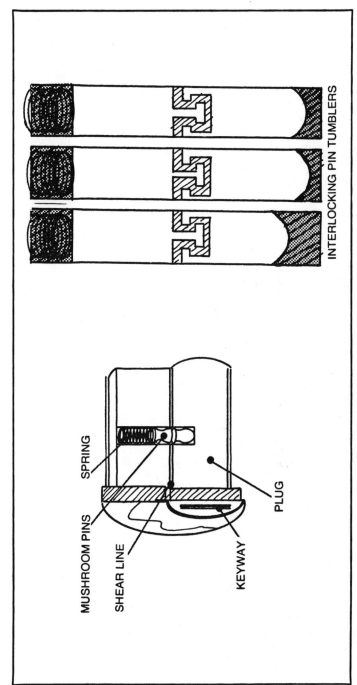

MUSHROOM PINS

SPRING

SHEAR LINE

KEYWAY

PLUG

INTERLOCKING PIN TUMBLERS

Fig. 2-18. Spool and interlocking pins as developed by the Corbin Lock Company.

Electromechanical locks, utilizing pushbuttons or magnet-impregnated plastic cards as keys, trip an electric circuit. There are locks which fire a hardened steel relocking device if the wrong key is inserted. With today's economy and the increase in burglaries, double-bolted and jimmy-proof locks with vertical bolts are enjoying an increase in popularity and sales.

Installation of the Primary Lockset

The *primary lockset* is the lock assembly normally installed by home builders. It is normally cylindrical and is commonly known as a *key-in-knob* lockset (Fig. 3-1).

There are three primary lock functions to be found on an entrance door. A simple spring latch provides the ultimate in convenience for the owner. When he leaves the premises, he closes the door behind him and it locks. The disadvantages of the spring latch are many. It can be easily defeated, often without detection. A common burglary practice is to force back the latch with a plastic card or thin shove knife (Fig. 3-2).

A *deadlatch* uses the same principle, but once the door is closed, the latch bolt is secure in the lock position and acts as a deadbolt. The deadlatch offers the same convenience as the spring latch but with more security (Fig. 3-3). Looking at the edge of the door, you will note a beveled bolt, characteristic of a regular spring latch. If it's a cylindrical design deadlatch, a small anti-pick button will project from the lock and rest alongside the latchbolt. In the mortise model, the deadlatch will take the shape of an additional beveled bolt, somewhat smaller than the latchbolt and placed higher on the lock body (Fig. 3-4).

In both the cylindrical and mortise designs, the latch bolt enters the strike plate to secure the door after it has closed. The deadlatch button, however, does not enter the strike, but is depressed into the lock housing when the door is closed. While

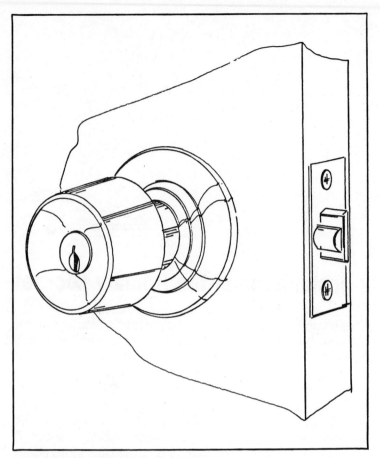

Fig. 3-1. Illustration of a key-in-knob lockset.

depressed, the deadlatch acts as a stop and secures the latchbolt in the extended position. The latchbolt can be operated by turning the inside knob, or using the proper key in the outside cylinder.

The deadlatch, properly installed, cannot be shove-knifed open. It can be defeated by jimmying or physically moving apart the door frame with a hydraulic jack.

Most primary locksets consist of a cylindrical lock with the locking mechanism built into the doorknob (sometimes referred to as a *key-in-knob lockset*). Consisting of a lock cylinder, a pair of door knobs and a spring latch, this type of lock is very popular and is used extensively throughout the world. Most modern American homes and office buildings have the key-in-knob lock as their primary lockset on exterior doors.

This lockset is comparatively easy to install and reasonably priced. The more secure locksets use pin tumbler locking mechanisms; the lower-cost, less secure locksets rely on disc tumbler mechanisms.

There are as many as a dozen companies manufacturing these locksets. While in outward appearance they are basically the same, there is considerable difference in construction details. Each assembles differently. Some are equipped with anti-pick latchbolts, others aren't. Some use steel, others brass, tin and plastic. In general, the more expensive locksets offer the best security and greater wear.

The key-in-knob portion of the lockset requires the drilling of a hole for the works of the lock (Fig. 3-5), and a second smaller right-angled hole through the edge of the door for the latchbolts.

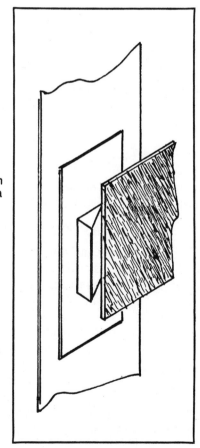

Fig. 3-2. Way in which a burglar can use a plastic card to force back a latch.

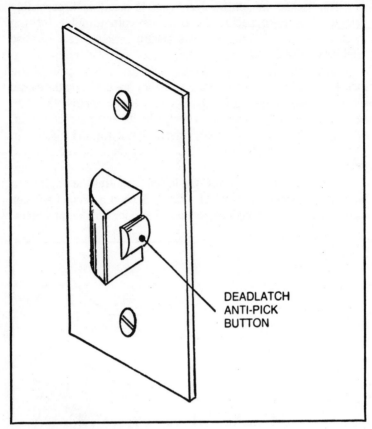

Fig. 3-3. Cylinder deadlatch.

INSTALLING THE LOCKSET

First read carefully the instructions pertaining to the particular lockset. Use the wood bit size recommended by the manufacturer. Then, using the template provided with the lockset, or taking your time with a ruler, mark out the holes accurately. Always keep in mind that it is easy to remove more if needed, but almost impossible to put it back without weakening security.

If you are installing more than one primary lockset in your home or office, it may pay to invest in a carpenter's bit or expansion bit; either can be purchased at a hardware or carpenter supply store.

Determine where the doorknob will be and draw a line at that point, approximately 36 to 38 inches above the floor. Then mark the door edge for centerline of door and mark center points. We will use the one-hole installation (Fig. 3-5).

Next, bore a 2-inch hole in the door face. Bore from both sides of the door to prevent splintering. Then bore a ⅞-inch hole in the door edge for the latchbolt assembly. Mortise for the latch face and install the latch (Fig. 3-6).

Install the exterior portion of the lock assembly through holes in the latch bolt assembly, and depress the latch while passing spindle and stems through latch holes (Fig. 3-7).

Then install the interior knob with the inside rose plate (Fig. 3-8) and push the rose tight against the door face. Line up the screw holes with stems, insert the screws and tighten until the lockset is firmly in place.

Finally, to install the strike plate, blacken the latchbolt and close the door. The strike opening should be approximately 36 to 38 inches from the floor. Place the strike plate on the door jamb,

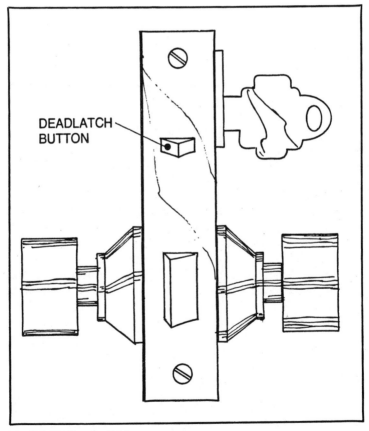

Fig. 3-4. Mortise deadlatch.

DEADLATCH BUTTON

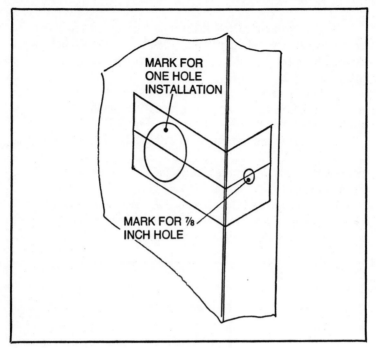

MARK FOR
ONE HOLE
INSTALLATION

MARK FOR ⅞
INCH HOLE

Fig. 3-5. Drilling a hole in a key-in-knob lockset.

centering the bolt opening with the blackened area made by the latchbolt; double check. The strike plate screw holes should be on the same vertical centerline as the latchbolt assembly screws. Cut out—(mortise) the jamb as needed and install the strike plate.

As illustrated above, most primary locksets are not difficult to install; but remember that each manufacturer provides his own template and instructions, and most will differ slightly in their installation requirements.

Maintenance problems with the key-in-knob locksets are relatively few. Because the knob is used most often, it wears faster than the rest of the locking mechanism. When the knobs wear, they can be replaced. The lock manufacturer's stock extra parts. At least annually, remove the screws and dust out the lockset's inner mechanism. Spray lightly with graphite, *not oil,* as oil collects dust and grime and may foul the lock. Reassemble.

INSTALLING THE MORTISE LOCK

Many office buildings, department stores and apartment houses utilize the mortise cylinder locks. There are variations, however, as when the house lock is key operated from the outside but can be locked from the inside with a thumb knob (Fig. 3-9). The more secure office or store lockset will normally use two key-operated cylinders; The second cylinder is installed on the inside of the door in place of the thumb knob. This enables the owner to control access by locking the door with the key and, if the door has glass panels, prevents the thief from gaining entry by cutting the glass and turning the thumb knob.

Installing the mortise lock requires care. First, read carefully the instructions provided with the lock. Next, determine where the door knob will be (normally 36 to 38 inches from the floor) and draw a line at that point. Then place the lock case flush with the door and

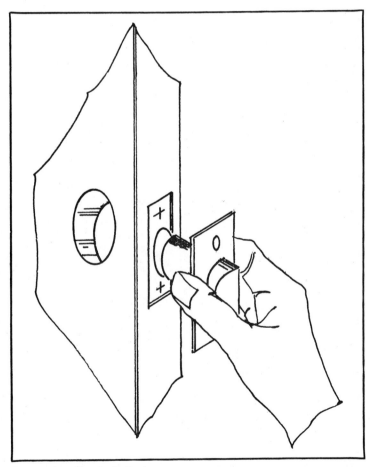

Fig. 3-6. Installing the lockset.

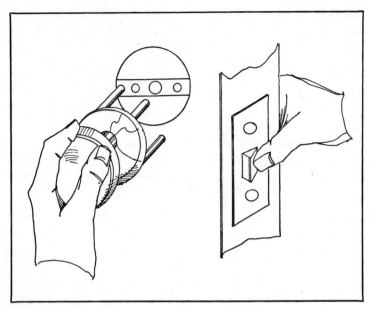

Fig. 3-7. Passing spindle and stems through latch holes.

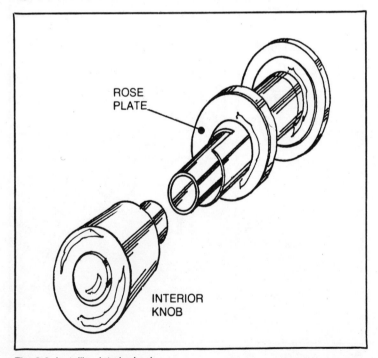

ROSE PLATE

INTERIOR KNOB

Fig. 3-8. Installing interior knob.

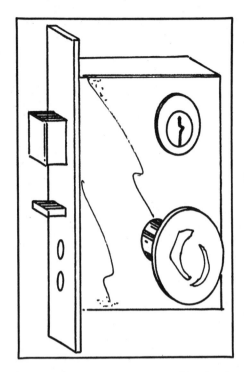

Fig. 3-9. A mortise lock with thumb knob variation.

using the doorknob line as a guide, line it up with the spindle hole in the lock case (Fig. 3-10), and then mark the lock cylinder hole. Next, draw a line around the edge of the door (Fig. 3-11). Then, using a wood bit equal in diameter to the lock casing, drill a series of holes straight in the center of the door edge. The line of holes should not be longer than the height of the lock case, nor deeper than its backset. Before inserting the lock case, take a wood chisel and gouge the cavity to receive the case. After double checking the cylinder and spindle holes, use a wood bit, 1-⅜ inch for the cylinder hole, and a similar size bit for the spindle hole. Tighten the lock case in place with two wood screws, and then assemble the lock trim (Fig. 3-12).

To install the strike plate for the mortise lock, close the door and mark the door frame. Then open the door and draw the proper lines (Fig. 3-13). To check your work, blacken the bolts with a pencil and close the door, letting the bolts strike the wood. The mortise strike should be installed so that the plate is flush with the surface of the door jamb, but no deeper. Take sufficient time to insure a snug fit. Otherwise, the repeated openings and closings of the door will soon loosen it.

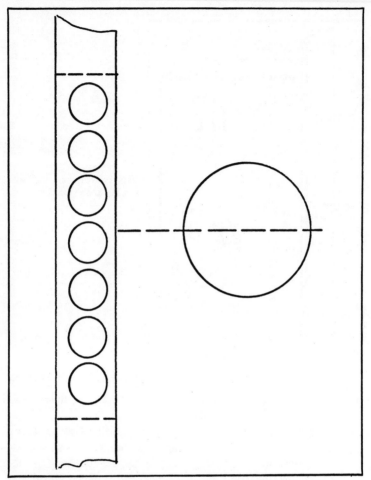

Fig. 3-10. Installing the mortise lock: using door knob line as a guide.

DETERMINING THE HAND OF THE DOOR

When installing locks, you must always check to insure that the beveled edge of the latch bolt faces in the correct direction; in other words, for proper installation, you must "determine the hand of the door." To do so, stand on the side of the door where you can see the hinges. If the opening edge of the door is at the left the door is a left hand door. If the opening edge of the door is at the right, it is a right hand door (Fig. 3-14).

If the lock is operated from either side of the door by a key, the hand of the lock should be the same as the door (right hand or left hand). However, if the lock can be operated by the key from the

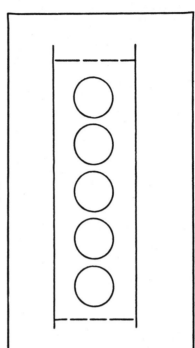

Fig. 3-11. Installing the mortise lock: drawing proper lines on door edge.

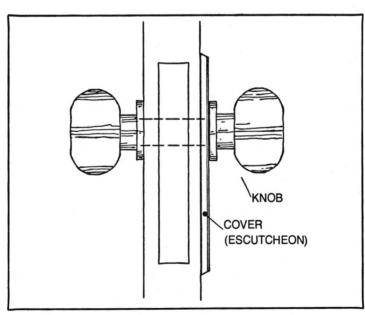

KNOB

COVER (ESCUTCHEON)

Fig. 3-12. Installing the mortise lock: tightening into place.

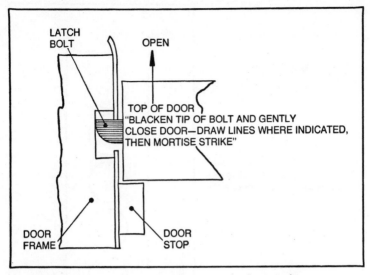

Fig. 3-13. Installing the strike plate: drawing proper lines on door.

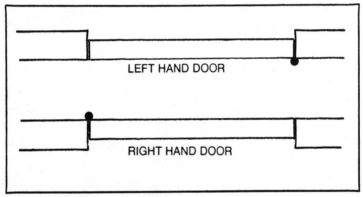

Fig. 3-14. Determining the hand of the door.

outside only, then the hand of the lock is determined from whether the latch bevel faces toward or away from the outside of the door. If the bevel of the latch faces toward the outside, it assumes the name of the hand of the door. But if the bevel faces away from the outside, it is called a *reverse bevel lock.*

Most locks today are made so that the latch bolt can be reversed. However, some tubular locks and some old fashioned mortise locks do not have reversible latchbolts. Always determine beforehand the direction the door swings and specify the bevel of the bolt when ordering the locks.

Installation of the
Night Latch/Auxiliary Lock

There are numerous lock models available today which can be classified as *auxiliary locks*. The most common is the *rim* or *surface-mounted* lock.

RIM LOCKS

The three basic designs of the rim lock are:

☐ *Common night latch:* least secure of the three, it works on the same principle as the primary lockset. It provides the convenience of self-locking when the owner slams the doors. Its primary weakness is its vulnerability to being shoved open with a knife.

☐ *Rim deadlatch:* again, it works on the same principle as the nightlatch, but it has the added feature of allowing the owner to move a button which sets the latch to deadlock. This prevents a would-be thief from shove-knifing the door open. Since this feature is activated only from inside the premises, this added measure of security applies only when the owner is in.

☐ *Rim deadlock:* a more secure lock, installed the same as those above. It comes equipped with rigid bolts measuring from one-half to two inches in length.

An added advantage of this lock is that the owner must lock it with his key; the bolt will not snap into place automatically (Fig. 4-1). Shove-knifing this lock is almost impossible, but the door can be jimmied open with a crowbar. A small jack can also be used to move the door frame the required distance to open the door.

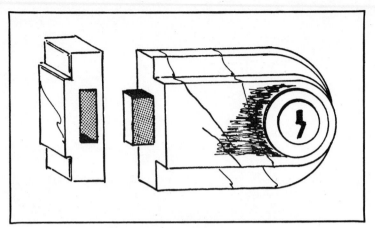

Fig. 4-1. The rim deadlock will not work without a key.

Since some burglars will go to any length to gain entry, the jimmy-proof lock was designed. The most popular jimmy-proof lock uses a bolt with two tubular shaped projections that drop vertically into the strike plate. Spreading the door frame with a crowbar or jack will not cause this bolt to come out of its strike plate (Fig. 4-2).

If the primary entrance has panes of glass positioned to allow access to the lock, it is recommended that you install a dual key cylinder lock on the door. This model has a normal keyhole on the outside of the door and a second keyhole, in lieu of a knob on the lock casing, inside. Once this lock is locked it can only be opened by a key. Should the burglar break the glass, he would still need the key to open the door.

The rim deadlatch auxiliary lock consists of three major parts: the lock housing, the key cylinder and the strike plate.

The rim auxiliary lock is easily installed. The first step is to bore the hole to accommodate the key cylinder; the standard size hole is 1¼ inches in diameter. The lock housing is placed flush with the edge of the door (Fig. 4-3). Then take a pencil or nail to mark the mounting holes.

The center of the hole must be 2⅜ inches from the edge of the door when the lock is being installed on doors opening inward (Fig. 4-4).

The center of the hole must be 2½ inches from the wooden stop or frame when the lock is being installed on doors that open out (see Fig. 4-5).

After cutting the hole, brush away the sawdust, take the key cylinder with its brass collar and place it into the hole (Fig. 4-6), ensuring that you do not leave the key in the lock. A lock is *never* installed while the key is in the cylinder. Glance at the keyway and adjust the cylinder so the keyway is at the bottom. If installed upside down (keyway at the top), the weight of the pins is too great for the upper springs, eventually the springs will collapse, causing the pins to drop. You will be locked out.

You are now ready for the small metal plate and the two grooved machine screws. Take these parts and assemble them to the cylinder (Fig. 4-7). You will notice that the screws appear to be too long. This allows installation of the lock to various door thickness. The manufacturers solved this problem by grooving the screws. Take two pair of pliers, and, being careful not to damage the threads, break each screw off to the length needed.

Next, place the lock housing over the connecting bar of the key cylinder. If the bar is too long and does not permit the housing to fit flush against the door, break off the necessary amount from the connecting bar. Make your break at the scored lines which are provided for this purpose.

Next, while holding the lock housing in place, try the keys in the cylinder. If they work freely, then the lock is positioned cor-

Fig. 4-2. A popular jimmy-proof lock with two tubular shaped projections that drop vertically into the strike plate.

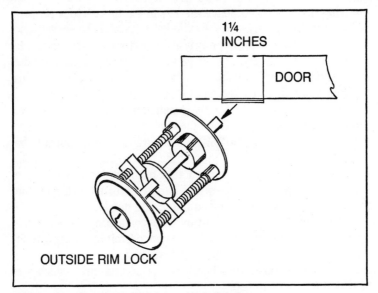

Fig. 4-3. The rim lock fits flush into a 1¼-inch hole.

rectly. If you experience a bind, move the housing up or down a fraction of an inch until the key turns freely.

After ensuring that both keys operate without binding, take a pencil and mark the screw holes through the holes in the lock housing. Remove the housing and make a starting hole at each mark with a nail or an awl, then replace the housing and fasten it with the wood screws provided. Do not overtighten the screws.

To install the strike plate, close the door and mark the location on the door frame. Most strike plates work for doors opening inward

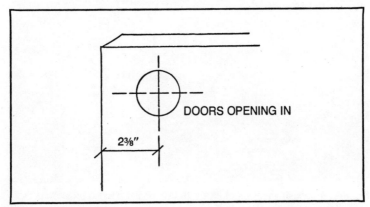

Fig. 4-4. Installing the rim deadlatch: proper distance of hole from door edge.

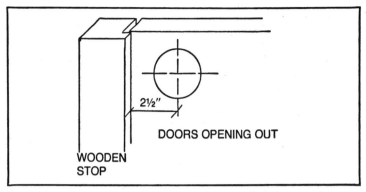

Fig. 4-5. Installing the rim deadlatch: proper distance of hole from wooden stop.

or outward. After centering the strike with the lock casing, mortise it into the door frame and fasten with wood screws.

TUBULAR DEADLOCKS

One major fault of many rim mounted deadlocks is that they are unsightly. In this situation, the tubular deadlock is the answer; it works exactly like the rim deadlock, but is mortised into the door like the key-in-knob set or cylindrical lockset.

For increased security, double key cylindrical models are available. They require a key to open the lock from either side and are recommended for doors containing glass or thin wooden panels.

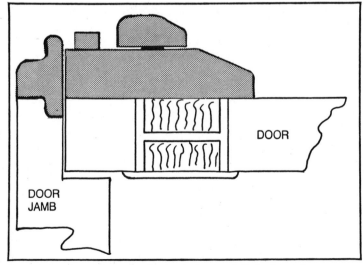

Fig. 4-6. Installing the rim deadlatch: placing key cylinder into hole.

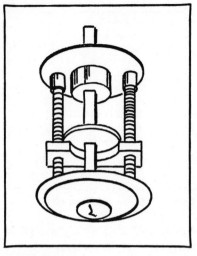

Fig. 4-7. Installing the rim deadlatch: assembling metal plate and screws.

A burglar who breaks the glass can release this lock by turning a thumb knob.

Tubular locks can be purchased with standard ⅝ inch deadbolts or long-throw deadbolts (from 1 to 2 inches). If the door fits poorly in the frame, or if an extra measure of security is required, the long-throw bolt is recommended.

The tubular deadbolt lock is installed in the same manner as the cylindrical (key in knob) lockset (see Chapter 3).

How to Identify, Repair, and Maintain Lever Locks

Although the outward appearance of the lever tumbler lock may change, all lever locks operate in a similar manner. The lever tumbler mechanism is relatively simple in appearance. Because each lever is hinged at a fixed point and held against a stop by pressure of a flat spring, and each lever also has a gate cut in it (all located at different places), the lever lock remains one of our most secure key-operated locks. It is used as the primary lock mechanism for safe deposit boxes in most banks.

Only one lever is visible (Fig. 5-1), but there may be any number of lever tumblers. When the proper key is inserted and turned, the notches on the key raise all the lever tumblers the required distance, lining all the gates up exactly opposite the post on the bolt. Turning the key still further retracts the bolt. But if the key is not the right one for the lock, causing even one gate not to line up properly, the lock cannot be opened. On the better lever tumbler locks (such as safe deposit boxes), serrations on the levers near the gate and on the post cause a binding if the improper key is tried. This lessens the lock's susceptibility to picking, adding to its overall security.

Figure 5-2 illustrates yet another lever design. The gate is an H-shaped hole cut into all the levers. Retraction of the deadbolt is possible only when all the levers are lined up with the center slot in the gate. When the bolt is in the locked position and the key is removed, the levers fall down. The deadbolt cannot be retracted.

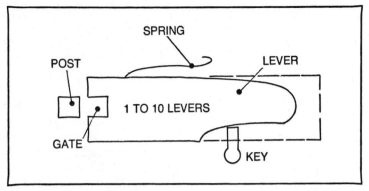

Fig. 5-1. Working principle of present day lever locks.

When the bolt is drawn back into the lock casing, the door opened, and the key removed, the levers fall down on the other side of the post. Whether locked or unlocked, the post is always snug in one of the two post positions. When the key is removed from the lock, the bolt can't move. This feature becomes necessary when it is desired to be able to remove the key in the unlocked position. Without it, the vibration of slamming a door might jar the bolt partway out, and then the key would be unable to either lock or unlock the door.

A similar concept can be found in the modern safe deposit box. When in the unlocked position, the user can't remove his key. In other words, the holder of the safe deposit box cannot return his box to bank vault storage and remove his key without first locking the box.

REPAIRING AND MAINTAINING LEVER TUMBLER LOCKS

The most frequent cause of trouble with lever locks is broken or disconnected lever springs. Individual levers are equipped with a spring. Should the spring break or fall out of its slot, the lever may jam, making the lock inoperable.

These springs are normally made of flat spring steel. They can slip out of place for many reasons: they have lost the tension necessary to remain in place; or perhaps they were too short to start with, were not properly formed, or have broken. Springs break because of impurities in the metal, improper tempering, or binding due to being inadvertently bent or kinked when constructed.

Take care, when installing new springs, to remove from the lock case all the broken pieces of the old spring; also be sure to keep the lever tumblers in their proper order. If the order is changed, the key will not work. When doing this, pay no attention to the numbers

56

stamped on the levers. These numbers do not refer to the position of the levers in the lock, but merely indicate the relative depth of the key cuts. An example: in a five-lever lock, you might find lever tumblers marked from 1 through 5; A shallow cut would be marked 1, a medium cut would be marked 3, and a deep cut would be marked 5.

The only sure way to keep the levers in their proper position is to remove them one at a time and place them in a line from left to right; then reverse the order when putting them back into the lock case.

Replacement spring stock should not be heavier than the original. Choose a spring thickness which will provide pressure equal to that originally used. Piano wire can be used for replacement springs.

The spring should be cut to the proper length and bent to the approximate shape before insertion into the spring slot. Ensure that this slot is clean and that the new spring is thin enough to fit snugly. On some lever tumbler locks the edges of the slot are turned slightly up and the spring can be held in place by taking a small ball-peen hammer and lightly tapping the edge. Care must be taken not to bend the lever tumbler or it will bind against other levers. After tapping, the lever, with its new spring, is returned to the lock together.

Chewed or worn levers are another problem. A steel key in use for a long time will eventually wear enough metal from the lever contact surfaces to interfere with the operation of the lock. The first indication of this is that the key becomes difficult to turn. A poor alternative is widening the lever gate with a file; doing this makes

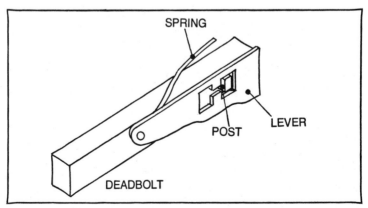

Fig. 5-2. Lever lock with an H-shaped gate.

the lock responsive to more keys and greatly reduces the lock's security.

The best method of repair is to clean up the rough spot with a fine grain file, and then file the lever a little thinner at the contact surface. This usually will flatten it enough to widen the lever where required.

A quick way to cure this ill would be to remove the lever entirely. But one must remember that reducing the number of levers in the lock proportionately reduces security.

Occasionally the post on which the lever tumbler pivots comes loose. The simplest repair is to peen with a hammer, riveting the fixed end solidly into the case.

Lever locks require a small amount of graphite every year or so. If some of the parts have begun to rust, take some light oil (preferably 3-in-1), and coat the parts. After a few hours, wipe dry with a clean, absorbent cloth. The coating of oil will soak through the rust and will prevent, to some degree any further rusting. No oil should be left on the levers. There is normally little clearance between the lever tumblers, and oil requires time to dry in which it can collect dust and grime, causing the levers to bind. Graphite, on the other hand, does not deteriorate with age and does not harden and thicken.

FITTING KEYS TO LEVER TUMBLER LOCKS

To fit a key to a lever tumbler lock, it is necessary to see inside the lock casing. This is simplified if the lock is fitted with a window or opening in the lock case just above where the post meets the gates on the lever tumblers. Remove the lock and use a length of stiff wire or a slim screwdriver to gently manipulate the levers one at a time until the bolt is in the thrown (locked) position. This is more easily accomplished if the lock case is held with the bolt pointing toward the ground. If the bolt is left in the withdrawn (open) position, the blank will be blocked from entering the keyway and you will be unable to proceed with fitting the key.

After securing the proper key blank, you must make the blank turn in the keyway so it can engage the lever tumblers. This means you must file a cut in the blank where it enters the keyway, permitting the key to rotate past the slot in the cover boss. Mark with a scribe on either side of the blank. Using a small thin flat file, make a cut straight down between the marks. Do not try to file too much metal at a stroke; file slowly until you reach the depth line.

Now insert the blank all the way into the keyway. Turn the blank to the right. Most likely you will not be able to turn it because the thin file is not as wide as the required width of the throat cut. Therefore, remove the blank and file a few strokes to the right side (tip side) until the blank turns freely in the keyway.

Before making the proper cuts, you must know exactly where to file the tumbler cuts. Take the key blank and blacken it with match or candle flame; be sure to darken the entire length of the blade, from the throat cut to the tip. Use a pair of pliers or vise-grips when blackening the blank as it quickly gets hot. Cover with only a thin layer or blackening. If you coat the blade excessively, the loose blackening will smear, or, worse, flake off in the tumblers. Magic Marker inks, nail polish, or other fast-drying preparations may be used successfully as alternates to blackening.

Next, the blank is inserted into the lock and given a slight turn. The position of the levers will be indicated by removal of the soot where the levers contact the blade.

Starting with the lever nearest the window (lever number one), remove the blank and carefully, a little at a time, file down on the blank until the lever has dropped to where its gate is in line with the post. This should not be done all at one time, but by repeated insertions, filing and examination. Then repeat the same routine for the second lever tumbler and so on, until the gates on all the levers form a straight line.

Care must be taken in filing the blank to ensure that the cuts are directly beneath their respective levers, and that no cut is so wide as to accept two levers, or so narrow as to catch a lever and thus stick. Always use a file approximately the same width as the required width of the lever cuts. You should then not have to widen the cuts. Just file straight down and ensure that your cuts are square and perpendicular. Proceed slowly, a little at a time.

If, after making the cut, you find that the lever is too thick to enter the tumbler cut, it will become necessary to widen the cut slightly. Be careful, however, not to make the cuts too wide. If the cut is too wide, more than one lever tumbler may be able to enter a single tumbler cut.

Try turning the key gently once all the gates appear to be in line with the post. If the key binds, look for the problem—it may be that each tumbler cut has to be a fraction deeper. If the key turns but seems to drag, reblacken the blade and unlock it again. The place it binds strongest will show as the brightest spot. A gentle flick with

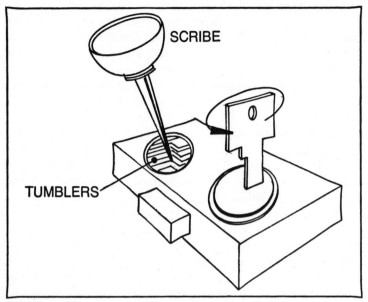

Fig. 5-3. Aligning gates with a scribe so key can be withdrawn.

the file should correct the trouble. Never force the key, as it may become trapped in the lock.

If, while cutting the key, you accidentally trap the blank, turn it in the locking direction as far as you can. Then, while holding the blank in this position, insert a scribe through the window (hole in lockcase) and align all the gates with the post (Fig. 5-3). Once the gates are aligned, the key will turn back and you can withdraw it from the lock.

Check the trapped blank for jam marks in either the middle or bottom tumbler cuts. If you see a jam mark (metal depressed) in either of the two cuts, file that cut slightly deeper. Try the key again in the lock. It should work smoothly.

Never cut a key from a metal softer than that composing the levers. Since most lever tumblers are constructed from brass, any harder metal will suffice. But avoid soft metal blanks. They break easily, wear too quickly and may provide real trouble.

There are a number of different types of lever tumbler locks in use throughout the world. Although they may look quite different in outward appearance, they all operate in the same manner.

The usual differences between lever locks will be the size and number of lever tumblers they contain. Naturally, the greater the number of levers, the greater the overall security of the lock.

The Disc Tumbler Lock

The *disc tumbler lock* was developed by the lock industry in recent years as an inexpensive lock substitute, ranking somewhere as far as security below the pin tumbler lock and lever locking mechanisms.

The disc lock resembles the pin tumbler lock in outward appearance, but in principle of operation it uses a much simpler mechanism and is far less costly to manufacture.

The disc tumbler operating principle, while essentially similar to the pin tumbler concept in operation, is very different in construction features. Picture a number of discs placed face-to-face in chambers in a plug or core. Each disc is pierced with a square hole called a key slot. These discs are equal in overall size, but the key slots, while identical in width, differ in depth.

The depth of the slot, commonly called the slot location, determines whether the tumbler requires a shallow cut or a deep cut in the key blank in order to line up at the shear line. The upper portion of the slot establishes the depth value for that disc tumbler.

The disc tumbler lock key looks almost identical to the pin tumbler key, except for its compact size. When the key is inserted, it goes through the key slot holes in each disc, lifting them to the aligned position. The key will then turn. The discs are properly aligned with the circumference of the plug, and the plug can rotate freely in the shell to operate the cam or locking bolt (Fig. 6-1).

To identify a disc tumbler lock, one merely has to glance into the keyway of the lock. Every disc lock has a series (usually five in

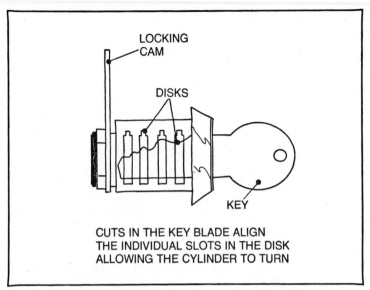

LOCKING
CAM

DISKS

KEY

CUTS IN THE KEY BLADE ALIGN
THE INDIVIDUAL SLOTS IN THE DISK
ALLOWING THE CYLINDER TO TURN

Fig. 6-1. The disc tumbler lock.

number) of chambers formed in the plug. The chambers house the disc with their springs. Each chamber is assigned a spacing number for ready identification, and these numbers are standard throughout the lock industry. The chamber nearest the keyway is the first spacing. The second is the second spacing, and so forth.

Looking into the keyway, one notes a flat disc riding somewhere between the bottom and the middle of the keyhole (Fig. 6-2). This flat disc identifies the lock as a disc tumbler.

The disc lock is frequently used as a less expensive substitute for the more secure pin tumbler lock in which security is not as pressing. The best disc tumbler locks are nowhere near as secure as the better pin tumbler locks.

The disc lock is normally found in use on cheaper cabinets and desks, and as padlocks, some automobile locks, locking handles, and inexpensive cylindrical door locks.

DISC TUMBLER LOCK CONSTRUCTION

Most disc tumbler locks consist of four major parts. The most important part of a disc tumbler lock is the plug, which contains a series of chambers passing right through the plug. An average disc tumbler lock has five chambers housing five flat disc tumblers, and a spring which exerts constant pressure, thus forcing the disc down so the lower portion of the disc extends through the plug into the lower locking groove of the shell or cylinder. This action locks the plug firmly in place and the plug cannot rotate in the shell.

The second part of the disc tumbler is the shell (or cylinder), which houses the plug; the third part is the cam (or lock bolt); which performs the locking function; the fourth major part is the retainer clip (or screw) which holds the lock assembly together. (Figure 6-3 illustrates the four major parts of a disc tumbler lock.)

Since most disc tumbler locks are produced with the manufacturer's eye to the selling price, greater tolerances are allowed in producing the locks, which goes along with the cheaper and softer metals. The end result is an inexpensive locking device providing a minimum degree of security.

The number of possible key changes for a disc tumbler lock is, in theory, limited to just over one thousand, but in actual manufacturing this is further reduced to between 250 and 500. Discounting the loss of security from looser fitting parts, the disc lock has only about one-tenth the security of a pin tumbler lock.

On some disc tumbler locks, the plug (or core) is held in place by a retainer clip. This rather simple device is standard on the cheaper model disc locks. By having the proper key, or by picking the lock to rotate the cylinder, thus freeing the plug from the shell, one can insert a small stiff wire through a hole in the front of the plug (Fig. 6-4) to depress the retainer. The plug can then be removed.

REPAIR AND MAINTENANCE OF THE DISC TUMBLER LOCK

Springs are the most common cause of trouble in the disc tumbler lock. Springs break or jump out of place, sometimes wedging between the discs and the chamber wall; when the key is inserted and the plug rotated, the lock jams or the key is damaged.

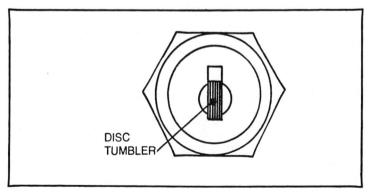

Fig. 6-2. Flat disc identifies lock as disc tumbler.

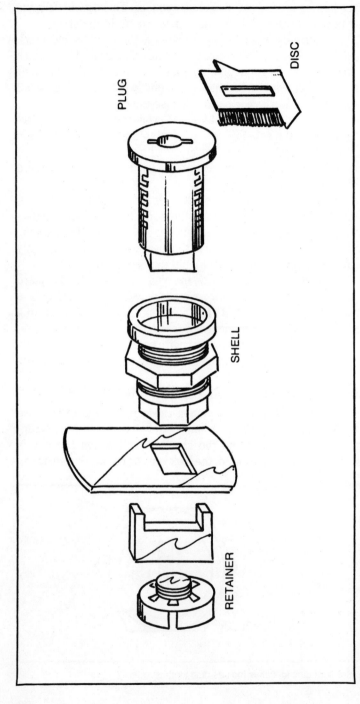

DISC

PLUG

SHELL

RETAINER

Fig. 6-3. Four major parts of the disc tumbler lock.

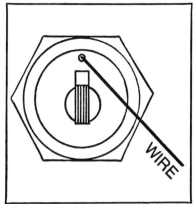

Fig. 6-4. Inserting stiff wire into disc tumbler lock.

A broken spring must be replaced. Springs that have slipped out of position must be put back, or if bent or weakened they should be replaced to prevent later problems. But first, some effort should be expended to determine the cause of spring movement. There is no point in replacing a spring only to have it jump free at a later time. Examine the other springs to see how they are held in place. It may well be that the disc spring wing is broken or worn so badly that the spring slips form side to side. In this case, the disc must also be replaced.

If the necessary disc or springs are not readily available from local locksmiths, one can simply remove the defective disc and/or spring. Remember that if the spring is removed, the disc must be removed and vice versa, otherwise the lock could jam. Remember also that this weakens the security of the lock because less resistance to picking remains. A bent or otherwise damaged disc should never be reassembled in the lock; it should be replaced or, at the least, removed.

When repairing the disc tumbler lock, place the lock keyway on the working surface and remove the retainer clip, screw or nut which holds the locking mechanism to the rear of the assembly. Then separate the plug from the shell; don't be concerned about the discs flying out. These are held in the plug by metal projections located at the top or bottom of the discs.

Place the plug in a small vise and remove each disc one at a time, starting with spacing number one which is nearest the keyway. After checking the disc and spacing for wear or obstructions, replace it before removing the next one. Note that it is important that the discs are returned in their correct order, otherwise the original key will not operate the lock.

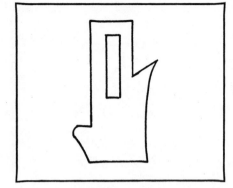

Fig. 6-5. Key slot high in disc.

If a damaged disc is simply removed, the original key will still operate the lock; but if a new disc is inserted without regard to the original depth of the key slot of the defective disc, a new key must be made.

FITTING A KEY TO THE DISC LOCK

As explained earlier, the key slot in the disc tumbler lock varies with each depth. When the key slot appears high in the disc the upper portion (top edge of slot) will be short (Fig. 6-5). The corresponding cut on the key blank must be shallow in order to align the disc flush with the circumference of the plug and free it from the shell. When the slot is positioned low in the disc, the upper portion will be long (Fig. 6-6). The corresponding cut in the key must be deep so that the disc clears the shell.

The overall length of the disc tumbler's upper section determines the depth value for that tumbler. Most manufacturers use

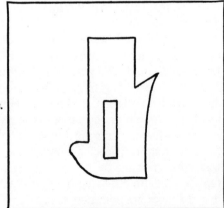

Fig. 6-6. Key slot low in disc.

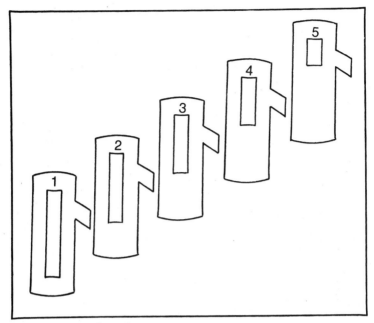

Fig. 6-7. Tumbler disc values.

only five depths. The disc with the shortest upper section is a number five depth. The disc with the longest is number one (Fig. 6-7).

The shorter the top edge of the disc key slot, the lower the key cut depth value. For example: number one would require a shallow cut; the longer the top edge of the key slot, the higher the key cut depth value, so that number five would require a deep cut (Fig. 6-8).

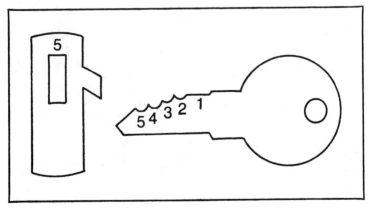

Fig. 6-8. A deep cut is necessary for number five disc.

Using a bright light, look into the keyhole of the disc lock and, with the aid of a small stiff wire about three inches long, you can readily determine what key depth is required to align each disc so the plug will rotate freely.

On most disc tumbler locks, the manufacturer will position a number four or five depth near the keyhole to block the view of the discs in back. Most locks—again for economy—do not use all available depths. The same depth disc may appear in two or more chambers.

The stiff wire is used as a reader to check the disc in back. It can also be used to count the number of discs in the lock. Since a lock may contain anywhere from one to six spacings, knowing the number helps to determine the proper key blank and eliminates the possibility of inadvertently omitting a key cut.

Hold the lock with the keyway tilted slightly downward (Fig. 6-9). Insert the wire in the keyway fully to the back of the lock and apply pressure down against the discs. This will force all the discs

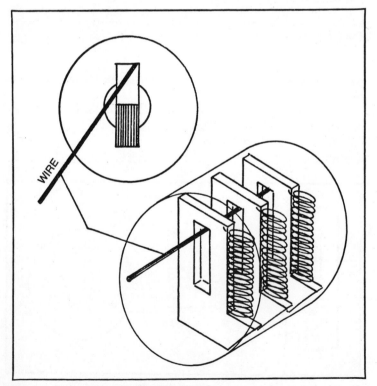

Fig. 6-9. Lock with keyway tilted downward.

Fig. 6-10. Using ward or rib to ascertain disc depth.

into the shell. Slowly withdraw the wire so that you can see and hear each disc click as it returns back into the keyway. You will see from one to six discs and will hear the same number of clicks as each returns to the "at rest" position. This determines the number of corresponding key cuts.

Next, replace the wire in the keyway and again depress all the discs. Slowly withdraw the wire until the last (front) disc clicks into the keyway. Then, using the keyway warding or rib (Fig. 6-10), you can ascertain the depth of that disc (Fig. 6-11 illustrates a number five depth as it rests in relation to the warding). Withdraw the wire slowly, reading all the remaining discs as each clicks into the keyway. Figure 6-12 illustrates a key with corresponding depth cuts of one to five. Always read the lock several times to be sure of the proper cuts before making the key.

Now that we know the depth cuts required to operate the lock, it becomes necessary to file these depth cuts in the key blank.

As an example, let's say you want to cut a key for a five disc tumbler lock with a depth reading of 1-2-3-4-5. These depths will vary in increments of twenty five thousandths of an inch (.025). Taking a section of stiff cardboard, and knowing that the precise

69

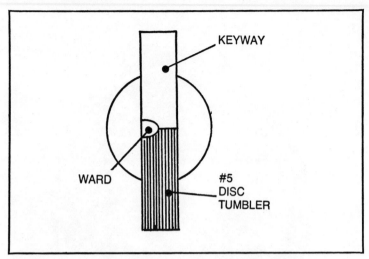

Fig. 6-11. Number five depth as it rests in relation to the warding.

difference in each depth is .025 inches, we can make the depth gauge (Fig. 6-13).

Next take the key blank and lay it on the spacing diagram (Fig. 6-14). Mark the blank with a sharp pencil where the lines indicate. These marks locate the positions where the disc tumbler will rest upon the blade when the key blank is fully inserted into the lock. The marks must be accurate or the cuts won't fit directly below the discs and the key won't work. Place the blank in a vise and, using a three-cornered file, faintly mark the blank where indicated by the pencil line.

Using the spacing diagram, widen the cut to the proper angle. Then, with the depth gauge, file the cut slowly until the notch in the gauge fits the cut snugly. File the cuts for the other spacings in the same manner.

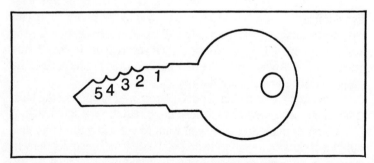

Fig. 6-12. A key with corresponding depth cuts of one to five.

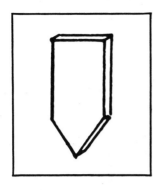

Fig. 6-13. Depth gauge.

When you finish filing the key, check the operation of the key in the plug. If it fails to operate, remove the key and determine if the cuts are correctly spaced. Are they filed to the correct depth and is the angle of each cut accurate?

Another, faster, method can be employed if the disc tumbler plug can be removed from the locking mechanism. First take the blank and, using the spacing diagram, carefully mark where the disc tumblers will rest upon the blade of the key. Place the blank in the vise and slide the plug all the way on. The blank will lift the discs, causing them to protrude through the plug. Each disc that extends beyond the plug casing requires a tumbler cut in the blank sufficient to drop it flush with the circumference of the plug (Fig. 6-15).

The required depth of each cut depends on how far the upper section of its disc extends out of the plug. The cut to align a disc that is just slightly out of the plug will be shallow. A cut to align a disc protruding much further out will be deeper.

Starting at the spacing closest to the bow of the key, file a slight cut in the blade. Use the angle gauge to check the angle of the cut as you widen it.

Filing only a slight amount each time, work with one disc until it is flush with the circumference of the plug when the key is inserted. Follow this same procedure until all the discs are even

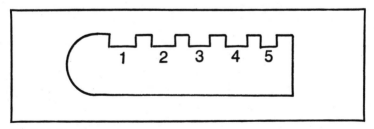

Fig. 6-14. Spacing gauge.

71

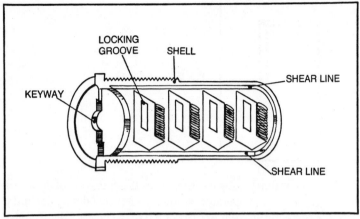

Fig. 6-15. Dropping the blank flush with the circumference of the plug.

with the outer edge of the plug. When finished, replace the plug back into the lock shell and insert the newly cut key. It should turn easily. Reassemble the lock with its cam or locking bolt and retaining device.

To duplicate the original key, take the proper blank along with the original, and clamp the two together with a pair of vise grips or on a small vise. Then take a match and blacken the exposed areas on the blade (Fig. 6-16).

Next, remove the original key. Carefully and slowly file with the three-cornered file to just before the bottom of the blackened area. Remove the key and check the cuts against the original. Continue to file until all cuts correspond.

Finally, take the new key and insert it in the keyway. It should rotate easily.

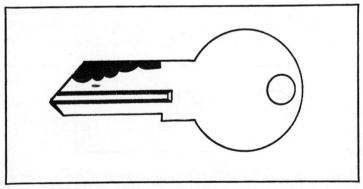

Fig. 6-16. Blackening exposed areas of the blade.

DISC LOCK MAINTENANCE

The disc tumbler lock requires only minor lubrication every few years. Oil should never be used because of the tolerances allowed in manufacturing which can cause dust and grime to collect between the shell and plug, or between the disc and its respective chambers, jamming the lock. A few drops of graphite will suffice.

If the lock operation becomes stiff, try graphite. If this doesn't help, check the key for wear. It may be worn so badly that force is necessary to align the disc tumblers in order to operate the lock.

As a last resort, disassemble the lock and check for built-up grime caused by the weather or by use of oil. Clean, reassemble and spray with a little graphite.

The Pin Tumbler Lock

The *pin tumbler* locking mechanism is the most popular type of lock in use throughout the world today. The greatest asset of the pin tumbler lock, aside from its burglar resistance, is its adaptability to all types of locking situations. The pin tumbler can be found on everything from a bicycle lock to money boxes.

To identify the pin tumbler lock, one notes that there are numerous keyways typical of this lock. These different shaped keyways (Fig. 7-1), along with the round pins or tumblers operating in a cylindrical plug, and the almost countless number of tumbler variations, provide the lock's security. Looking in the keyway you see a round pin protruding upward (Fig. 7-2).

A pin tumbler lock may contain as few as three or as many as eight spacings, but most locks are constructed with five or six pin spacings. Each spacing in the pin tumbler cylinder will contain a spring, an upper pin (driver pin), and a lower pin (Fig. 7-3).

The upper pin, as noted, is flat on both ends. Lower pins are flat on top and round on the bottom. The round or tapered bottom permits the lower pins to seat fully within the cuts of the key, it also permits them to ride easily over the separations between key cuts as the key is inserted or removed from the lock.

The combined action of the spring, upper pin and lower pin within the plug and lock cylinder performs the locking function (Fig. 7-4). Without the action of these pins, a screwdriver or stiff piece of wire could be used to rotate the plug in the shell and operate the lock.

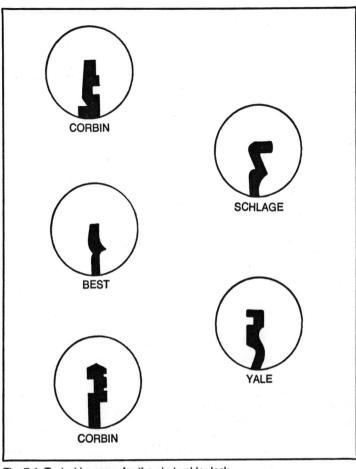

Fig. 7-1. Typical keyways for the pin tumbler lock.

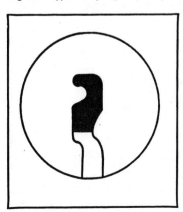

Fig. 7-2. Round pin in keyway of pin tumbler lock.

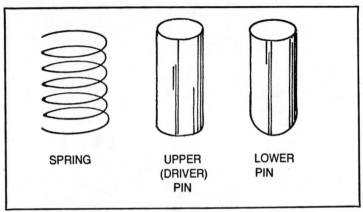

Fig. 7-3. Contents of spacing in pin tumbler lock.

The spring and the upper and lower pins as installed in the lock are in the "at rest" position. Spring pressure against the upper pins has forced it, and the lower pins, into the plug. Figure 7-5 illustrates how the upper pin extends down past the shear line, locking the plug and shell together. They thus cannot turn.

To turn the plug and operate the lock, the set of pins must be lifted just high enough to be able to separate the upper and lower pins at the shear line. The lower pins must be raised until their top is flush with the top of the plug. This in turn will lift the upper pin fully into the upper chamber. When this occurs, the pins will be aligned at the shear line and the plug will turn, operating the locking bolt or cam.

When the upper and lower pins do not separate at the shear line—meeting just above or below this line—the plug will not turn. In a five-pin tumbler lock, the five pins must be at the correct height

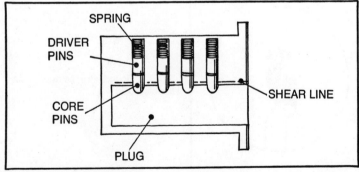

Fig. 7-4. The pin tumbler lock.

SHEAR LINE

Fig. 7-5. How the upper pin locks plug and shell together.

to allow plug rotation. Figure 7-6 illustrates a pin tumbler lock with the proper key in place. Note that the upper and lower pins separate at a common line called the shear line (the circumference separation between the plug and the shell).

Sometimes a manufacturer will use steel ball bearings riding atop the lower pins; they will rotate freely as the key is inserted and withdrawn from the lock. These balls do not alter the operation of the lock, but will frustrate any attempt to defeat the lock through picking. The ball bearing near the keyhole can be seen by looking into the keyway (Fig. 7-7).

To operate a pin tumbler lock, we have observed that the key must first fit the keyway and must be of the correct length to be able to raise the pins to the correct height.

Manufacturing tolerances permit the key bittings (cuts in the key) to be any of ten different heights, ranging from zero to nine. Today pin tumbler mechanisms are used in padlocks, cabinet locks, key-in-knob locks, automobiles and utility locks (cam locks).

REPAIR AND MAINTENANCE OF PIN TUMBLER LOCKS

The majority of problems with the pin tumbler lock are confined to the cylinder itself. It may be something as simple as a loose

Fig. 7-6. A pin tumbler lock with proper key in place.

cylinder, which allows the whole cylinder to rotate when the key is inserted and turned instead of holding it in place while the plug turns inside it.

If it's a rim cylinder, remove the lock casing from the door and tighten the two retaining screws (bolts) holding the cylinder in place. If it's a mortised lock, tighten the set screw located in the lock's face-plate.

Quite often the correct key is hard to insert in the lock and it may also be difficult to remove. First check the key for rough spots or a bent blade. If the key cannot be inserted and removed easily from the lock, there may be a piece of broken key or some other foreign matter lodged in the far end of the keyway.

Also, if the key becomes wedged partway in, requiring considerable effort for its removal, someone may have tried to force the lock with a screwdriver, thus mashing the pins and narrowing the keyway.

Remove the foreign material or broken key with a piece of stiff wire or construct a key extractor (Fig. 7-8) by using a length of fret or coping saw blade. A mashed keyway can sometimes be opened with a small flat file, but since this eliminates the warding and opens the lock to other key blanks, it's a temporary measure and the cylinder should be replaced.

Sometimes the correct key will not operate the lock. The cause may be nothing more than a loose cam at the rear of the lock. Remove the cylinder and tighten the cam screws. At this time, examine the cam for wear. If the cam is tightly in place, but there is still play in the plug, remove the cam and very slightly file the shoulders of the plug down. This will make the plug and cam fit snugly and prevent problems when the key is withdrawn.

Fig. 7-7. An example of steel ball bearings riding atop lower pins.

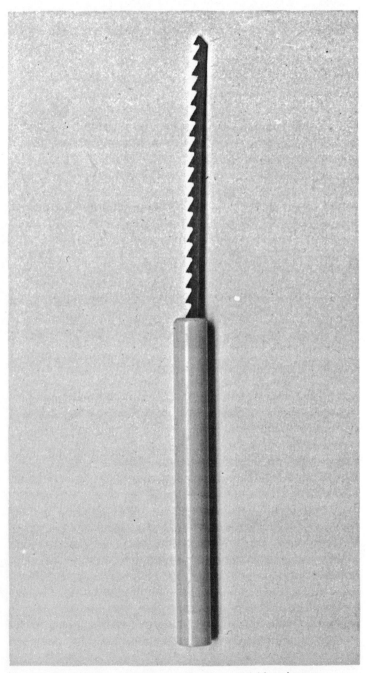

Fig. 7-8. Key extractor used to remove foreign material from keyway.

If the lock operation is still not satisfactory, soak the cylinder in gasoline. It may be gummed up with an excess of oil. After the gasoline has thoroughly dried out, a puff of graphite should suffice.

PLUG REMOVAL

Removing a pin tumbler plug is easy if you have the correct key and the proper tools. First, remove the key cylinder from the lock mechanism. Place the cylinder face down and remove the two screws at the rear of the plug (these screws fasten the plug retainer in place). Next, remove the retainer and connecting bar or cam (Fig. 7-9). Observe caution when removing the plug retainer screws. *Never* hold the cylinder in your hand. Always place the cylinder face down on a flat surface and hold firmly in place while removing the screws.

Make a couple of storage containers from paper cups or plastic bowls to contain small parts. You can make a pin tray from styrofoam or balsa wood (Fig. 7-10).

Next, place the cylinder carefully in a small vise and tighten the vise gently. Too much pressure will crush the cylinder shell. Insert the correct key all the way into the plug and *hold it there.* Do not pull back on the key: remember that the retainer is off. The pins are at the shear line and the plug can easily fall out, releasing the upper pins and springs. Turn the key approximately one fourth turn to the right and center the plug retainer screw hole in line directly below the upper pin chambers. While still holding the key in the plug, pick up the universal follower or follow-through (a rod of metal or wood slightly smaller in diameter than the plug); (Fig.7-11), and place it directly behind the plug (Fig. 7-12). This will create a smooth and continuous surface for the upper pins. Keep the follower pressed

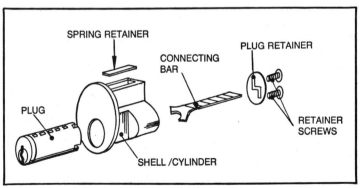

Fig. 7-9. Removing connecting bar and retainer.

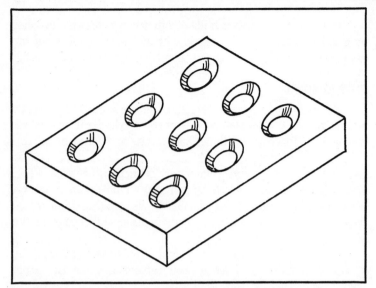

Fig. 7-10. Carve or cut rounded ½-inch-deep holes in material used.

against the rear of the plug. While offering slight resistance to the plug with one hand, begin to slide the follower forward through the cylinder shell with the other. The plug will come out the front without permitting any upper pins to drop out of their chambers. *Do not* separate the plug and follower at any time; this will create a gap and allow the upper pins to fall. To complete this operation, push the plug all the way out until the follower completely replaces it in the cylinder.

Another method to remove a cylinder without access to a cylinder cover cap or proper key is "shimming". Shimming is the procedure of inserting a thin metal strip (approximately .0015 of an inch) between the plug and the cylinder shell at the shear line in order to separate the upper and lower pins without the aid of the correct key. Once the retainer and retainer screws have been removed, place the cylinder gently in a small vise. Insert the shim at the top of the plug directly beneath the upper pin chambers. Be sure

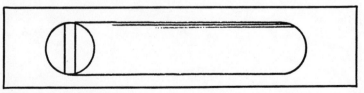

Fig. 7-11. A universal follower.

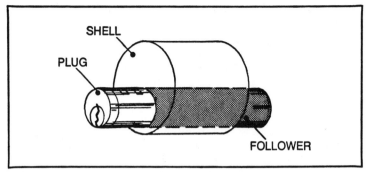

Fig. 7-12. The universal follower in use.

the shim is between the plug and cylinder shell and not in the keyway.

Advance the shim until it rests against the upper pins in the last (back) chamber. Do not force it beyond this point. To do so will cause the shim to buckle. A key blank is now inserted fully into the keyway. Maintain a slight amount of forward pressure on the shim as you slowly withdraw the blank from the plug. As the blank is withdrawn, the set of pins in the last chamber will begin to slide down the tip of the blank. At the exact point where the upper and lower pins separate at the shear line, your pressure on the shim will cause it to slip between them.

It may be necessary to work the blank very slightly back and forth against the pins a few times until they are lifted to the proper level where the shim can slide between them. But once the shim has passed between the pins, don't force the blank back in as it will damage the shim and possibly the pins.

Once the pins in the last (back) chamber have been separated, you will feel it slip easily forward until it is again obstructed, this time by the pins in the next chamber. Follow the same procedure with the blank and shim until the upper and lower pins in the remaining chambers have been separated in the same manner.

Once the pins in each chamber have been separated by the shim, the tip of the blank may be used to turn the plug. Remember, if the plug uses a connecting bar instead of a cam, turn the plug to the right far enough so that the connecting bar clearance slot won't be aligned under the upper pin chambers. To complete this procedure, remove the shim and key blank. Separate the plug from the shell by pushing out the plug with the follower. Shimming a cylinder correctly is a matter of practice and timing. By timing the in and out movements of the key blank while applying forward pressure on the shim, you will align the upper and lower pins at the shear line long enough for the shim to slip between them.

Occasionally, poorly fitted or worn keys shift in relation to the pins in the plug. To correct this, either disassemble the lock and file down or change the pins, or blacken the blade of the key and, after locating the high spots, replace the necessary pins.

When the correct key is inserted but requires considerable force to make the plug turn, it's possible a short pin has worn so much that it's at an angle in its chamber. Another possibility is that a broken spring may have fallen between the pins. In both cases, disassemble the lock and replace the spring or pins affected.

At times a key will operate easily when the door is open but has to be forced if the door is closed. In all probability, the door has moved or the door frame has warped or settled; this causes the deadbolt to bind against the edge of the strike plate when the door is closed. This binding causes the key to turn with difficulty. Correct by raising the door and/or file the strike plate as needed.

If the latchbolt binds or sticks, check first for paint or the possibility that someone has tried to force the door. Other problem areas are dirt or other foreign matter in the lock, broken springs, or a worn latch/bolt. Determine the problem. If dirt or foreign material is found in the lock, clean with gasoline or solvent and replace the lock. Broken springs can be replaced, but if the latchbolt is greatly worn, it must be replaced.

Sometimes when the proper key is inserted and turned you will notice a slight drag or sticking of the plug before it responds. Normally this is caused by a worn key. But often-times a key that has not been perfectly cut for that lock will cause this reaction. The effects might be improved by blackening the key blade and filing. If the key is badly worn, it is best to simply replace it.

Never lubricate a pin tumbler lock with oil. Dust and grime will build up in the lock and soon cause jamming. Pin tumbler locks should always be lubricated with powdered or flaked graphite. Use the graphite sparingly. It's extremely messy and too much can cause problems later. After applying the graphite, work it through the lock by running the key quickly back and forth. The easy way to use graphite is from the squeeze tube available at most hardware stores.

HOW TO EXTRACT A BROKEN KEY

First ascertain that the plug is lined up with the upper pin chambers. This is important because the broken portion of the key

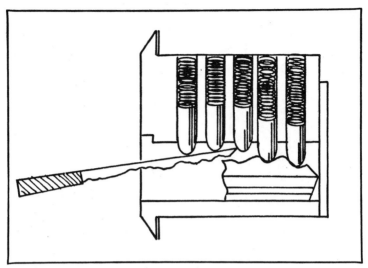

Fig. 7-13. Inserting key extractor into keyway.

will not move unless the lower pins are free to move into the upper chambers as the key is withdrawn.

Take the key extractor (coping or fret saw blade; Fig. 7-8), and insert it all the way into the keyway until you can grasp the first cut in the key segment (Fig. 7-13). Then raise the blade parallel with the top of the keyway, thereby lifting the forward pins, and slowly pull the blade out; the key will come out.

If the broken portion of the key is protruding partway out or is flush with the keyway, take the extractor along one of the grooves of the key, making the saw teeth bite into the solid part of the key segment and then slide outward (Fig. 7-14).

FITTING A KEY TO A PIN TUMBLER LOCK

Duplicate the original key by hand. First purchase the correct key blank from a locksmith (most department stores now also sell blanks). Or if you have a key blank that fits the keyway and can be further cut to match the original key, place the two side by side in a small vise, blacken the areas to be cut and file with a small file until the key operates the lock. Care must be exercised to keep from cutting too deeply. Make shallow cuts at first and try the new key from time to time until it operates the lock.

Fitting a key to a pin tumbler plug: Place the plug with lower pins removed into a a small vise. Insert a key blank. Next, mark the center for each cut in the key, insuring that the blank is fully seated

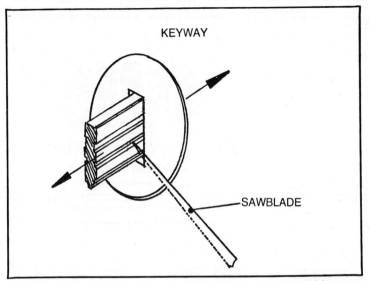

KEYWAY

SAWBLADE

Fig. 7-14. Using extractor when key is broken partway out or flush with keyway.

in the plug. Take a prick punch or nail and insert it into the first chamber. Tap the punch lightly with a hammer. This will cause a slight indentation, marking the exact center for your first cut. Do not hit the punch hard. Follow this same procedure to mark the centers for the remaining cuts. Tapping may cause the blade of the key to back out of the plug. To prevent this, be sure to push the blank fully into the plug prior to marking each chamber.

Next, remove the blank from the plug and tighten it in a vise. Faintly notch the blank at each mark with a three-cornered file, (Fig. 7-15).

Place the blank back into the plug and check for centering of the notches through the chambers. Replace the first pin in chamber one of the plug (the chamber nearest the keyway). Notice how the blank lifts the pin high above the top of the plug. The cut in the blank will have to be deepened to bring the pin down flush with the shear line (circumference of the plug).

Slide the plug off the blank and gently file the first cut slightly deeper. Use the key gauge to check proper angle of cut. Again, slide the plug all the way on the key. The pin should have dropped closer to the top of the plug. Remove the plug and continue to file, making shallow cuts and checking constantly until the pin rests flush with the top of the plug. Repeat the same procedure for chambers two through five (common pin tumbler lock).

After all pins are installed flushed to the top of the plug, remove the key and plug from the vise and, with it turned slightly to the right, slowly insert it back into the cylinder. Remember to keep the plug and follower pressed tightly together until the plug completely replaces the follower in the cylinder.

Rotate the plug back to center so that upper and lower chambers are in alignment. Place your thumb firmly against the face of the plug near the keyway and gently remove the key; place the

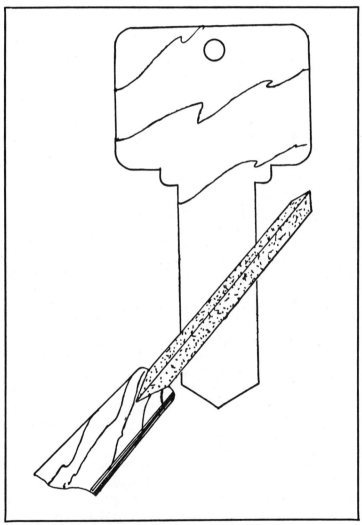

Fig. 7-15. Notching the blank with a file.

cylinder face down on a flat surface and replace the retainer and screws. Insert the key into the plug; it should turn freely and withdraw easily.

Fitting a key by the impression method: Keys can be fitted to pin tumbler locks without picking open or removing the lock. The novice will find the procedure complicated and time consuming, but once you have mastered this method you will find it a fairly quick way in which to make a key.

The impression method is based on the fact that the lower pins of the lock will leave a mark or impression when you exert pressure on the blank while it is in the keyway.

Equipped with the correct blank, take a pair of Vise Grip pliers or C clamp and attach to the bow (handle) of the key. Insert the blank fully into the keyway and twist and rock firmly; avoid bending the blank. Use just enough pressure to obtain the impression marks without putting undue strain on the blank.

Remove the blank from the lock and examine the top of the blank carefully, holding the blank under the light. To get a clean view of the mark, roll the blank very slowly. When the light strikes the top of the blade at the correct angle, you will see several clear impression marks.

The blank can be blackened to better see the impression. Take care not to coat the blade excessively to prevent the soot from flaking off and entering the lock. Also, the blackened area can wear off as the tumblers slide on the blade when the blank is inserted, leaving an uneven pattern which can blur the impressions. Do not use graphite on a lock when you intend to fit a key by impression.

If you experience difficulty in seeing the marks after working the key in the lock and rolling it to make the impressions appear, it is possible you are working under too bright a light. The bright light may cause too much reflection. Use a smaller bulb or move away from the light to check the blade.

To help obtain a clear set of impression marks you can file the blade of the blank thinner. A thin blade does not resist the pressure of the pins so much and therefore a deeper, more pronounced mark will result. Furthermore a thinner blade will wiggle more easily in the keyway.

At times not all the pins will mark the blade. Normally the longer pins in the lock will mark first. Therefore, you may see only two or three marks and the others will not appear until you have filed the cuts and permitted the longer pins to drop down into the keyway.

Using a round Swiss file, make a shallow cut where you see the sharpest or deepest impression mark. Work two or three strokes of the file at a time, very slowly. File only one mark at a time.

After each filing, insert the blank into the lock and try for another impression mark. Stay with the first cut until no further mark is obtained.

When you no longer get an impression where you are filing, start to file the next sharpest impression mark. Eventually the blank is filed until every pin reaches the shear line, allowing the newly cut key to turn, freely operating the lock.

A variation of the impression method is to insert the blank fully into the plug. Then slip a steel rod through the key ring hole in the bow of the key and apply, turning pressure to the steel rod. Now tap gently on upper and lower edges of the bow. These impressions are normally clean and distinct.

How to Improve the Security of Your Home or Office

Begin to improve your home or office security by *always* locking your doors. Locks are the first line of defense against the would-be burglar. Never leave doors unlocked when departing, even if it is just for a few minutes. Control distribution of your keys and change the locks if you lose them.

Challenge unfamiliar faces in your neighborhood or office area. Report all suspicious actions or persons to the police. Make a checklist of the present security devices employed in your home or office. Look at these barriers to unlawful entry as if you were the thief. Will your present devices slow him down? Or does your home invite entry by its obvious lack of good locks and the presence of vulnerable areas?

While it is almost impossible to stop the professional from entering your premises, you can discourage the amateurs by making the entry as difficult and time-consuming as possible.

Many burglaries are done on an impulse, on a dare, or as a lark. Some are by persons normally honest, but in need of cash. Upon finding a door or window open or unlocked, he will enter and take whatever is handy and easily convertible to cash. If the door resists his efforts, if the windows are locked and if he is challenged, he will more than likely move on.

A recent major burglar study revealed that neighborhood self-protection programs, designed to alert the citizen to the steps he can take to safeguard his home or property, have been very successful. Such things as home security checklists, improved lighting, adequate locks, and even guards and alarm systems are being used to deter the would-be thief.

HOW TO IMPROVE THE SECURITY OF DOORS AND WINDOWS

Often the difficulties encountered with an insecure locking mechanism can be traced to improperly fitted doors, inferior door hardware, a sagging frame and changes made to the door after installation of the lock assembly. Any of these causes can prevent the lock from performing its primary function: securing your home or office from unauthorized entry. At times the latchbolt will not hold or will only seat properly when the door is slammed. Because the building industry allows certain shortcuts in constructing the rough frames for doors and windows, a natural action—such as contraction or expansion caused by interior or exterior temperature changes or shifting and settling foundations—can cause a latchbolt or even a deadbolt to be inoperative.

Security of one's residence or office cannot depend entirely on the primary locks installed. For centuries it has been a standard building practice to mount interior and exterior doors on pre-finished frames that are installed over a rough frame. This practice makes good business sense, for only the outer finished frame need be constructed with any degree of accuracy, and the inner frame can be constructed of unfinished lumber assembled within tolerances of plus or minus a half inch or more.

More and more wood or metal outer frames are being constructed at the factory, then equipped with hinges, strike plate and door stop, and shipped to the construction site with a fitted door mounted in the frame.

The door frame is then slipped over the rough frame and nailed or bolted into place. Various sizes of wood shoring is used to space the frame the correct distance from the inner frame and to make its sides plane and vertical. Generally there may be anything from a half inch to over one inch space between the outer frame and the support frame. Lengths of wood or plastic trim is then affixed to the edge of the frame to hide this space.

The frame hardware (normally hinges and strike plate) are recessed into the finished frame and held in place by short wood screws. Under normal operation, these screws should hold fast for years with little or no tightening. But when the screws have loosened through abuse or lack of preventive maintenance, the screw holes may become destroyed.

To correct this, simply remove the screws and fill each hole with glue and small wood particles or sawdust (or use epoxy cement). For a split wood frame first clean, then fill with epoxy and clamp together until the cement has dried.

Replace the short wood screws for added strength and security. Remove the wood trim and back the frame with a length of board of the correct thickness and length. The length and thickness is important to keep the frame properly aligned with the door. Once the backup board is securely in place, drill holes through the frame and the back board to the thickness of the larger screws. It may be necessary to enlarge the holes in the hinges or strike plate before a larger screw can be fitted.

Insure that you bevel the new holes, allowing the larger screws to be flush. If they protrude, the door will not close properly.

Solid core doors are generally used today for exterior passage. They are the best. Some residences and office complexes still have hollow core or glass panel doors. Wood or glass panel doors should be made more secure before affixing good locksets. Back glass panels with sturdy and attractive metal grilles. They are available at most hardware or department stores. Use carriage bolts or lag bolts to secure them to the solid portion of the door. For aluminum or metal doors, the glass panels can be covered in the same manner, using metal tap screws to secure them in place (Fig. 8-1).

Hollow paneled wooden doors can be reinforced at a minimal cost with half-inch to one-inch thick plywood mounted to the inside of the door.

To prevent the would-be burglar from gaining entry through a gap between the door and the jamb by shove-knifing the bolt or using a crowbar, a latch guard should be installed. This device is constructed from angle iron (heavy steel) folded lengthwise at a ninety degree angle or shaped to form a T-bar (Fig. 8-2), and measuring from six to twelve inches in length; it is fastened to the edge of the door by lag bolts and when the door is closed it hides the latchbolt.

For increased security a two sectioned latch guard is affixed to the outer and inner portion of the door at the latchbolt area (Fig. 8-3). and held in place by two smooth-head bolts fastened to the inside by lock washers and nuts.

The latch guard protects the opening between the door and the frame at the proximity of the latchbolt. When the door is closed, a latch guard provides complete protection against jimmying and shove-knifing without modifying the lock or its installation. Often

METAL
GRILL

Fig. 8-1. Metal grilles affixed to door.

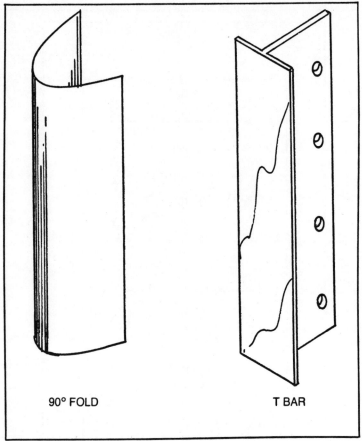

Fig. 8-2. Typical latch guards.

the presence of a latch guard is enough to persuade the would-be burglar to look elsewhere for an easier target.

Once the doors and the primary locksets have been made more secure, it is time to back it up with a secondary/auxiliary deadbolt lock. The preceding chapter on auxiliary locksets discussed the merits of the deadbolt to improve one's overall security. The more secure deadbolt locks are equipped with vertical bolts and are almost jimmyproof (Fig. 8-4). Their weakest point is the mounting of the strike plate. Improve your strike plate security by backing up the door frame; use longer wood screws to secure the strike deeply into the studs.

Another popular method of forced entry is to wrench or pull the key cylinder out of the lock assembly, and then operate the lock manually from the outside to open the door. To prevent this, install a cylinder guard, available at most locksmith shops and hardware stores. A cylinder guard plate shrouds the outer key cylinder with a drill-resistant steel cover (Fig. 8-5). A small round hole in the plate permits the owner to insert his key and operate the lock. The plate is held in place by carriage bolts (smooth head) or by bolts that pass through the door from the outside, secured with nuts and

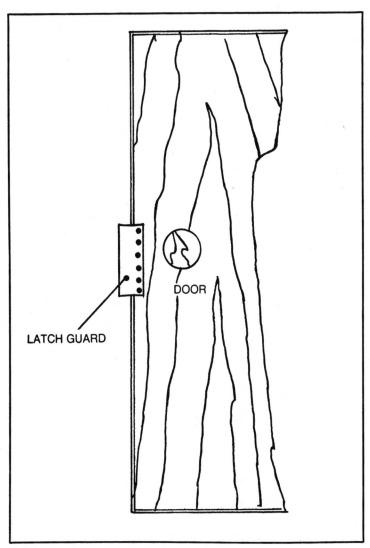

Fig. 8-3. Installation of latch guard.

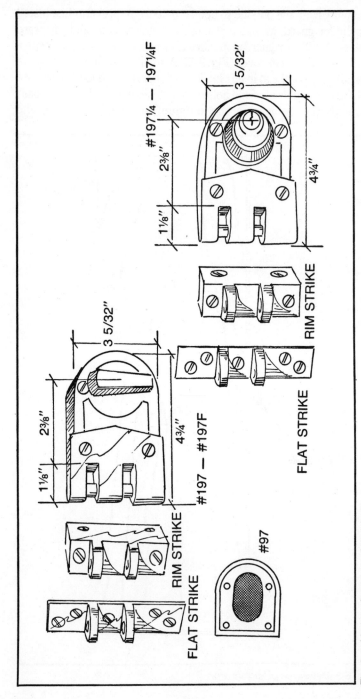

Fig. 8-4. Various jimmyproof deadbolts.

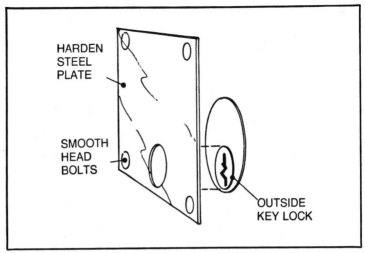

HARDEN
STEEL
PLATE

SMOOTH
HEAD
BOLTS

OUTSIDE
KEY LOCK

Fig. 8-5. Cylinder guard plate.

lockwashers on the inside. While it detracts from the appearance, it is also very effective.

For office doors with glass panels that can be easily broken, a dual cylinder lock should be installed. Glass paneled doors can be reglazed using Lexan, an impregnable transparent (space age) plastic that resists cutting or breaking. For one-quarter-inch thick sheets, the approximate cost is four to six dollars a square foot.

Exterior doors should be fitted with wide angle viewers (commonly called peepholes). Once installed, they should be used. They will be of little value if a member of the family continues to open the door to every friendly voice requesting audience. A convex mirror installed opposite the viewer will reveal anyone attempting to hide outside the range of the viewer.

Do not rely entirely on chain locks. Chains can be snapped or cut with little effort. While a chain lock permits the owner to open the door sufficiently to see and speak, the person outside can cut or forcibly rip the chain off the door frame. If he has the time and skill he can defeat it with a shove-knife or rubber band (Fig. 8-6). Remember: the chain lock was never intended to take the place of an auxiliary lockset. It's an accessory item designed to allow the resident to screen callers. If one is so inclined, he can construct a heavy duty chain lock using a length of 750 pound test chain, purchased from a hardware store. One merely forms a loop around the door knob and attaches the loop to the door frame and stud using a three to four inch lag bolt.

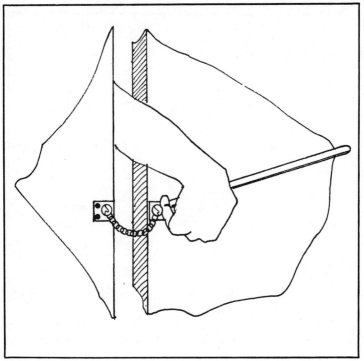

Fig. 8-6. How a burglar can shove-knife a chain lock.

To secure little-used doors, the brace lock will provide the greatest degree of protection. The lock body is installed flush to the door facing and the brace (a long steel bar) is anchored to the floor (Fig. 8-7).

In the locked position, the top portion of the bar is fastened in a slot in the lock body. Pressure on the door forces the bar more securely into the lock. The brace lock is not meant to be convenient or pleasant to look at. However, it is by far the most secure design yet invented, and is used extensively in high crime areas where property owners are willing to sacrifice convenience and beauty for peace of mind. The one limitation is that the brace locks only work on doors which open into the premises.

Sliding glass doors are very common in most newer homes. Some businesses have one or more sliding doors installed. Because most sliding doors are at the rear of the building, they are easy prey to illegal entry. Unfortunately, most owners secure their sliding glass doors using only the simple cam-latching device built into the door by the manufacturer. This provides *no* security.

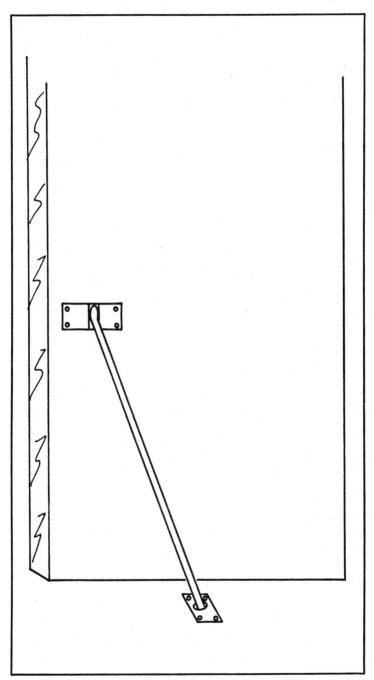

Fig. 8-7. Brace lock in place.

Various locks have been developed to secure sliding glass doors and are available at most hardware stores or locksmith shops. These differ in design, and can be found for use on single or double sliding sections. Some are simple jamming mechanisms, others are key operated.

An easy way to secure a single-section sliding glass door is to use a bar lock. The bar is hinged at the bottom side of the stationary glass door, approximately two inches from the bottom track and hidden from outside view, and swings into a support mount on the sliding door. When the bar is in position, the door will not move. The bar can be pivoted up into the hinge bracket, thus freeing the door for use. An expedient lock can be constructed out of a broom handle or piece of wood cut to the exact length and placed in the track on which the door slides, preventing the door from being pushed open even if the cam-lock is forced.

Because a sliding glass door is lifted into position when installed and, therefore, can be lifted from the track and removed, it is advisable to install large head screws into the top of the door frame track at both ends and the middle. Adjust the screws so the door just barely clears them when it is operated.

The most secure lock for sliding glass doors is a vertical deadlock engaging bolts in reinforced holes in the floor and ceiling. Also, you can use a key operated lock, attached to the inside track door, and then drill a hole in the outside track door to receive the bolt. When the key lock is operated, it depresses the lock bolt into the hole, effectively securing the door.

IMPROVING WINDOW SECURITY

All windows in your home or office should be securely latched. You must not neglect the windows. You can be sure the burglars will make an attempt through an unprotected window. There are three basic types of windows. Double hung windows are the most common. Most windows use sash locks, which do not lock, but merely engage; they prevent the upper window from falling down due to gravity, and the lower window from being forced up from the outside. They can also prevent a child on the inside from raising the window, these devices are *not* locks, since they are not key-operated and *not* secure. While they may stop the child, they will hardly deter a highly motivated thief. The sash lock can easily be opened with a table knife and then with a screwdriver slipped between the window sections (Fig. 8-8).

Many homeowners will go to considerable effort to improve

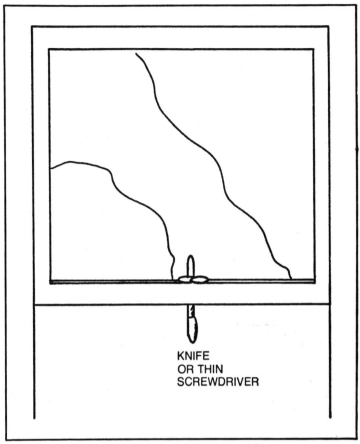

KNIFE
OR THIN
SCREWDRIVER

Fig. 8-8. How the sash lock can be opened from outside.

the security of their home by installing solid-core doors with good locksets, and then fail to even consider their windows as a means of easy access.

When planning a home or office, you should ensure that all first-floor windows are high off the ground and are small. But if you already have a home or office, there are still many ways to secure the windows.

Mounting decorative metal grilles on deadsash (non-moving) windows with lag bolts is one effective way to prevent access. Metal screen guards are available through locksmith shops; they come in various designs. They may have a bell or siren attached to an alarm trap which will go off when an entry is attempted through the window. These may be battery or electrically operated.

The simplest method to secure an occasionally-used window requires a hole to be drilled through the side of the window and into the frame. A stove bolt is then slipped through the hole, holding the window fast.

Windows not needed for ventilation should be sealed permanently. Others may be fitted with key-operated locks, offering freedom of operation and maximum security. To permanently seal a window not in use, simply drive a large nail or screw into the window track so it can't be raised far enough to allow an intruder to slip through.

Casement windows are generally considered most secure. To make a susceptible casement window more secure, you can replace the crank lever lock with a key-operated one available from many hardware stores or locksmith shops. In most cases, simple removal of the window crank will deter most would-be burglars.

The same information for securing the home applies to protecting an office complex or commercial establishment. But businesses need even greater security and there are numerous ways to provide the needed protection.

High-security pick-resistant lock cylinders can be installed in place of the conventional pin tumbler cylinders supplied with your locks. Where duplication of keys is a concern or when the owner just wants a key cylinder that is difficult to pick open, a high-security cylinder is in order.

A high-security cylinder is of sophisticated design and uses keys that restrict unauthorized duplication. It may have from five to seven pin tumblers equipped with spool (mushroom shaped) pins or interlocking pins (Figs. 8-9 and 8-10). The proper key must be cut at the required angle to correspond to the shape of the pins. In the Medeco cylinder, the key lifts the pins to the shear line, and twists them to align all the slots with a spring-loaded sidebar in the cylinder shell. Then, with the shear line clear and the sidebar able to retract, the key will turn, retracting the cam or locking bar. Corbin's

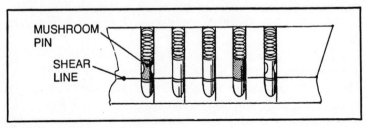

Fig. 8-9. High-security cylinder.

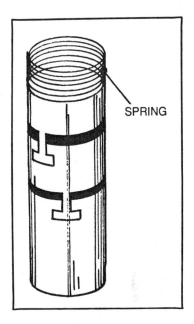

SPRING

Fig. 8-10. Interlocking pins of sophisticated design.

high-security interlocking pin tumbler cylinder uses a specially cut
key available only from the manufacturer.

The cylinder plug has six circular grooves central to the six pin
chamber holes (Fig. 8-11). The upper and lower pins have slots for
interlocking set at a 20 degree angle to the centerline of the tip
(rounded point) of the lower pin. To open the lock, the pins must be
aligned by the proper key with bitting cut on a corresponding 20
degree angle. This allows the interlocking end of the upper pins to
fit into the six grooves in the cylinder plug. The plug can now be
rotated since the key has slots corresponding to the circular
grooves in the plug. The interlocking parts engage and disengage
smoothly because the angle of the key bitting maintains a rigid
relation of the pin position to the grooves in the cylinder plug. It is
also for this reason that, without such a key, the pins are almost
impossible to align by any other means.

Sargent Lock Company's Keso twelve pin tumbler locking
mechanism is yet another high-security cylinder that is unconven-
tional in design. Rather than having the normal five or six pin
tumblers in a single row, a total of twelve pins are used with four in
each of three separate rows (Fig. 8-12). The keys are quite different. Instead of cuts being made in the top of the blade, milled
indentations are made on the top and side to align all three sets of
pins to the shear line simultaneously. Again, Keso keys cannot be

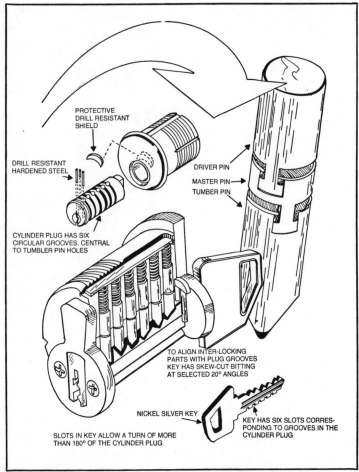

PROTECTIVE DRILL RESISTANT SHIELD

DRILL RESISTANT HARDENED STEEL

DRIVER PIN

MASTER PIN
TUMBER PIN

CYLINDER PLUG HAS SIX CIRCULAR GROOVES, CENTRAL TO TUMBLER PIN HOLES

TO ALIGN INTER-LOCKING PARTS WITH PLUG GROOVES KEY HAS SKEW-CUT BITTING AT SELECTED 20° ANGLES

NICKEL SILVER KEY

KEY HAS SIX SLOTS CORRES-PONDING TO GROOVES IN THE CYLINDER PLUG

SLOTS IN KEY ALLOW A TURN OF MORE THAN 180° OF THE CYLINDER PLUG

Fig. 8-11. Cylinder plug with its six circular grooves and pin chamber holes.

duplicated on conventional key machines, but must be ordered from the manufacturer. The sale of key blanks is restricted to listed customers. These factors, plus the pick-resistant design, make the Keso an extremely high-security key cylinder.

HOW TO PROTECT OFFICE EQUIPMENT

Most of today's modern offices are outfitted with the latest in time-saving equipment such as electric typewriters, tape recorders, calculators and copying machines. These same time-saving machines are very popular with burglars who can dispose of them quickly, usually at a fraction of their original value. Even with good locks on the doors and windows to the office, sometimes thousands of dollars worth of office equipment can still fall prey to a thief.

To stop the burglar, individually lock *each* piece of equipment so it cannot be removed from the premises. Use a maximum security office equipment lock. This lock is designed to anchor the equipment to the desk or surface by a hardened steel bolt that passes through the lock base and threads into the chassis. When the pick-resistant lock mechanism is inserted in the base and locked, access to the bolt is blocked and the equipment is secure.

While a single lock is normally sufficient to protect most typewriters and small pieces of office equipment, for further protection you can install a hardened steel locking bar with the office security equipment lock. This will provide two point security up to distances of twelve inches using one lock. To secure large, bulky equipment such as televisions or reproducing machines where the distance exceeds twelve inches, use two or more equipment locks.

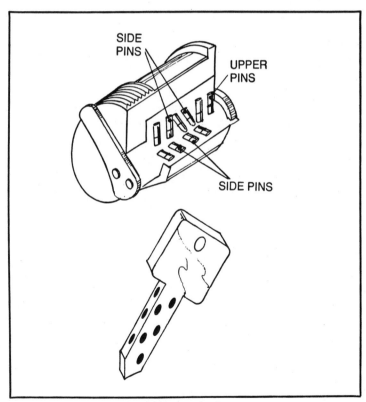

Fig. 8-12. Twelve pin tumbler locking mechanism.

An economical way to protect records or account receivables stored in filing cabinets is to install a filing cabinet bar lock. Installation is simple. The bar can be purchased in kits, containing all the necessary hardware and instructions, at most office supply houses. The bars are available in different sizes to fit almost every type of cabinet.

With a little time and patience you can construct a filing cabinet bar lock with a length of strap (angle) iron approximately six inches longer than the height of the cabinet. Bend at the top and drill a 1-½ inch slot about two inches from the end. Take a padlock hasp and drill two holes through the top of the cabinet to correspond with the slot in the bar. Next, affix the hasp to the top of the cabinet using two carriage bolts with lock washers and nuts. At the bottom of the cabinet, directly below the last drawer, attach an eyelet constructed with hardened steel to receive the metal bar. Then, insert the locking bar through the drawer handles and into the eyelet, attaching a padlock to the hasp at the top, securing the cabinet.

No matter what security precautions you take, everything will still depend on how fully the habit of security becomes a part of your everyday pattern of living. The finest locks or alarms will do little good if the door key is hanging on a nail near the door or under a mat, of if a duplicate has been made for part-time help. Security demands a price—not so much in money as in a constant alertness and adherence to specific security guidelines, even if they require some daily inconveniences and changes in long-established routines.

Keying Alike and Master Keying

More and more locks installed in todays homes use the same key to operate both front and rear doors, as well as the garage. This is called "keying alike."

KEYING ALIKE

To do this—to *key alike* the various locks installed in a home or office—they must first all be of the same family. That is, the keyway in all the locks must correspond to the one key, and all the locks must be either pin or disc tumbler locksets, but not a mixture of both types. If all the keyways are alike and the locksets are of one type, e.g., pin tumbler, then one only needs to change the lower pins of the various plugs to fit the key and operate the locks.

If the cylinders are different from one another, then you must purchase new cylinders with identical keyways, available at most hardware stores or locksmith shops.

When replacing a cylinder, ensure that the new cylinder matches the old one in size. Therefore, when you purchase the replacement cylinders, keep the following in mind: most modern cylinders use the same diameters. Always check the threads closely on a mortise cylinder. While most mortise cylinders utilize a standard thread around their circumference, a few are not standard. So be sure to check the original cylinder against the one being purchased. To compare the threads, hold the cylinders together so the threads interlock. Similar threads will interlock perfectly, while different threads will not.

Mortise cylinders will vary in length. Check yours. Remember, you can always use a *longer* cylinder if you cannot match the original one, but you can seldom use a *shorter* one. When using a longer mortise cylinder, you can use an expansion cylinder ring between the lip of the cylinder shell and the door surface to fill up the empty space.

To replace a rim cylinder, first remove the lock case from the inside of the door. Then remove the two cylinder clamp screws and push the cylinder outward. After ensuring that you have the correct replacement cylinder, you are ready to reassemble the lock. Check the cylinder connecting bar to ensure that the original connecting bar and the replacement are similar. Never use a cylinder with a flat connecting bar to replace one with a square connecting bar. You will find, however, that the most common connecting bars are flat.

Connecting bars are attached to the plug in one of two ways: horizontal or vertical. Again, most manufacturers have designed their cylinder plugs to accept a universal plug retainer.

To change the angle of the connecting bar, merely remove the plug retainer and flip it over, then re-insert the connecting bar into the retainer at the required angle and fasten the retainer to the plug.

Keep in mind that, with keyed alike cylinders one key operates all the locks. This weakens security for the sake of convenience. If you should lose a key or have one stolen, *every* lock will have to be changed.

MASTER KEYING

While keying alike offers convenience, a much better and more secure system is *master keying.*

Master keying is the process of setting up a series of locks to be keyed differently for normal use by various persons, such as in an office building, hotel or home; yet it allows one authorized key (the master key) to operate a whole series of locks.

While such a system of keying locks is not practical for most home applications, master keyed systems are usually favored for commercial or institutional use. But for increased security, master keying can be applied for residential use.

While most lock families (warded, lever, disc or pin tumbler) can be master keyed, it is doubtful that master-keyed warded locks can be found today due to their simple construction and limited uses. Disc tumbler locks can be master keyed, but, aside from desk locks, they are seldom bothered with. One type of lever lock series which are master keyed are safety deposit boxes. Normally, the second lock on the box operates with a master key controlled by a bank official. In a small bank, one key may operate the second lock, while in a larger bank there may be several keys, each operating a group of locks.

Normally, pin tumbler locks are the ones most often master keyed. Pin tumbler locks are the most popular type of lock in use throughout the world. Pin tumbler locks are used on almost all the exterior doors in all countries. Because of this, it is most often the lock system to be master keyed.

There are numerous ways to master key a series of pin tumbler locks. One system requires different keyway widths to admit certain keys while excluding all others. The width and shape of the keyway determines what key will operate the lock. This is a simple form of master keying, allowing the holder of the thinnest key access to all the keyways, but preventing keys one size larger from entering all locks. This system offers little security, since, to gain access to all the locks, one merely has to shave the blade of the key to fit the keyway of all the locks.

Another method is the extended length, pin tumbler lock. It may come equipped with from six to ten pin chambers. To master key this system, one merely uses keys of different lengths. For example the first series of locks may be fitted with five pins positioned to the front of the lock. The second series of locks would be fitted with seven pins (again, beginning with the front pin chambers) so that the shorter key will not reach and cannot operate the seven pin tumbler lock. This master key system can be set up using any number of pins and key blade lengths. The longest key becomes the master key and, with the aid of a middle set of pins, called master pins or chips, allows the different keys to operate the lock. Master pins are not used in every pin chamber, but only in those chambers where the master key fails to lift the lower pins to the shear line (Fig. 9-1).

We have discussed two relatively simple master key systems, which may be adequate for a small business with two to five locks. But if one is considering master keying a motel or an apartment building to control access (and to allow an authorized person access in case of an emergency, such as a fire or water leakage), then a more complex master key system is needed.

The secret of master keying is that each master cylinder contains not only a set of upper and lower pins, but also a middle set of pins called the master pins (chips), which line up at the shear line when the master key is inserted. While it is not always necessary to

Fig. 9-1. An example of master pin use.

place a master pin in every chamber, the master pins must be used in those chambers where the master key fails to lift the lower pins to the shear line.

For example, to master key four cylinders to operate with individual keys, while allowing one master key to operate all four locks, you first must disassemble each cylinder so that you can insert the required master pins into the pin chambers.

For this you will need a following tool, assorted pins and chips and the individual keys for each lock. Using the key for each lock, you can separate the plug from the shell with little difficulty.

Take the first lock and insert the key. Remove the retainer and cam or connecting bar. Turn the key about one quarter of a turn to the right or left. Next, while firmly holding the key in that position, very carefully push the plug out from the rear with the follower pressed firmly against the plug.

After determining which is to be the master key, insert it into the plug. You will notice that a pin or pins may extend to or past the circumference of the plug, while others fall short of it. Keep in mind that the correct arrangement requires all the pins to reach the shear line (circumference of the plug).

Next find the correct master pins to completely fill the space in the chambers to the shear line. Some chambers may not require a master pin.

Remember the correct size master pin must reach the shear line but not extend above it. If the pin protrudes above the shear line, then it is too long. The master key will not operate the lock and it will jam up.

However, if the master pin falls short of the shear line, you will have another problem. The plug will insert easily back into the shell, but once you have reassembled it, the master key will not operate the lock. The too-short master pin will allow the upper pin to extend past the shear line, thereby locking the cylinder shell and plug together (Fig. 9-2 illustrates a proper master key and plug arrangement). Always remember, with the key inserted the pins must rise to the circumference of the plug, but no pins must extend beyond the circumference of the plug.

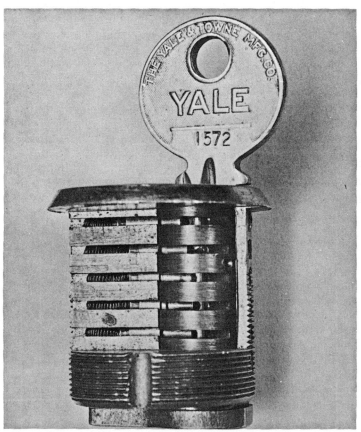

Fig. 9-2. When upper pin extends past shear line, cylinder shell and plug are locked together.

To reassemble the first cylinder, and after determining that all pin chambers are equipped with pins that stop at the shear line, you simply insert the plug without removing the master key in the shell. Be careful to hold the two together firmly while installing the retainer. Now turn the key to check the cylinder for smooth operation. Next, remove the master key from the cylinder and check the operation of the original key. Both should operate the cylinder smoothly.

To master key cylinders two through four, simply follow the same procedures. First insert the master key to determine where master pins are required (normally only one or two chambers). Again select the proper master pins to fill these chambers to the shear line and install them. Reassemble the cylinders and check both the master key and the original key for smooth operation. Your master key system is now complete. The master key will operate all four locksets, while the individual key for the first lock operates only the first lock and so on.

If any key fails to operate its original cylinder, disassemble the faulty cylinder and check to see that all lower pins rise to the shear line with the original and/or master key inserted. If they do not, rearrange them until they operate with either key.

The method of master keying just discussed is recommended for up to a maximum of ten locks. As a master keyed system multiplies and more pins are added, security is weakened, thereby increasing the chance of burglary and causing lock maintenance problems.

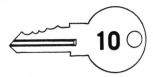

How to Gain Entry When Locked Out

Night after night, television drama demonstrates to us how both the good guys and the bad guys can manipulate intricate locks with little effort. A quick twist of the hand and the lock gives way; deftly inserted plastic cards placed between the door and the trim, and the door swings inward.

Thankfully, bypassing and picking a lock is not that easy. The simplest warded lock requires some knowledge of its inner workings and a certain amount of skill. The more complex the locking mechanism, the more knowledge, skill and time required to defeat it.

GENERAL HINTS

If you yourself are locked out, stay calm. Think. Determine the easy way in. Always, before attempting to pick a lock or forcing entry, check for an unlocked door or window. Also check the hinges or the clearance beween the door and frame. Often locks do not have to be picked open, nor property damaged by forcing entry.

Various techniques are known for opening doors and windows without picking the lock or harming the premises. And while all these "tricks of the trade" will not work in every situation, one or more will succeed in most situations. Remember to think clearly and keep calm.

If your check determines that the door opens outward and the hinges are affixed to the outside of the door, the simplest way through the door is by removing the hinges. On older buildings it is only necessary to use a screwdriver and a hammer to tap the hinge pin out. The door may then be removed from the frame without damage to either the door or the lock.

Many modern buildings today are equipped with a set hinge pin on doors opening outward. These have a set screw which is concealed when the door is closed. Frequently, because of sloppy workmanship, these screws have not been tightened, allowing removal of the hinge pins just as readily as that of a loose pin hinge. Even when you find a tightened hinge pin, it simply means you must apply a little more force. The set screws for most hinged pins do not penetrate to any great depth; consequently, the added applied force will usually release the pin.

Shove-knifing or shimming is likely the oldest method for opening a lock without a key and it is still useful in many situations. If a spring latchbolt holds the door in place and the bevel of the latch faces the outside, the latch can be sprung with a piece of flexible plastic, a thin screwdriver or a thin metal shove-knife (Fig. 10-1). Slide the knife between the door and the trim in the area of the lockset and feel for the latchbolt; then apply pressure to retract the bolt, freeing the door.

When the bevel on the latchbolt faces inward, you will need to make a curved or L-shaped shove-knife (Fig. 10-1) constructed from one-inch spring steel or sheet metal. To retract the spring latch, simply insert the shove-knife between the door and the trim near the area of the latchbolt, placing the L-shaped end of the knife behind the latchbolt; pull it toward you.

Sometimes a sharp ice pick will suffice for both spring latch positions. Determine the area of the latchbolt and push the pick firmly against the bolt while sliding the pick toward the door.

To defeat a rim mounted deadbolt lockset, place a hydraulic jack, with two six inch sections of two-by-fours, across the door frame. Turn the jack slowly, just enough to free the deadbolt. This is possible due to the tolerances allowed between the carpenter's rough frame and the prefabricated outer frame measuring up to an inch on either side, more than enough to defeat most deadbolt locks.

To defeat a safety chain lock where the door is unlocked but the chain prevents entry, simply take a rubber band and tie one end, leaving just enough to slide over the knob end of the safety chain. Next, take the other end of the rubber band and a thumb tack and, reaching around the door as far as possible, affix the rubber band to the door with the tack, then carefully, slowly, pull on the rubber

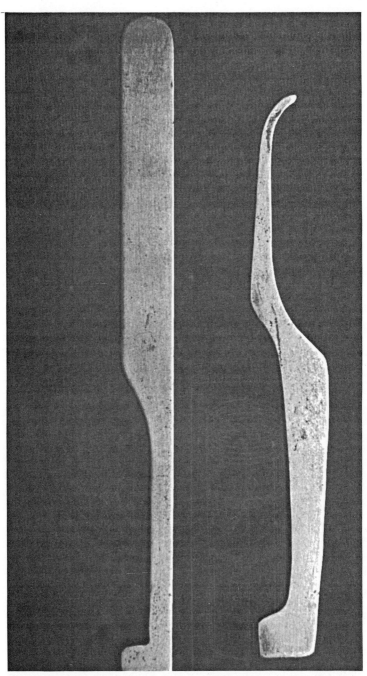

Fig. 10-1. Metal shove knife (left) and an L-shaped shove knife (right).

band, placing the tied end over the knob. Upon closing the door, the rubber band will pull the chain and release it (Fig. 8-6). Be careful when opening the door, since the chain is still hooked to the rubber band.

The key-operated safety chain lock can be defeated merely by taking a one inch spring steel shove-knife and pushing up to release the spring latch.

Most double hung windows are equipped with a cam-locking device called a sash latch. These are not locks and they are not secure. They merely prevent the upper window from sliding down due to gravity. To open the window from the outside, take a narrow putty knife or kitchen knife and slip it up between the window halves. Move it forward, catching the sash latch and pushing it open.

Metal casement windows require that you merely remove the putty from around the glass, then remove the glass and reach in and release the latch. While it's easy to remove the pane of glass from a metal frame because of the rigid outer surface, wood frames are much more fragile and must be handled carefully. Otherwise you may ruin the entire window.

Sliding glass doors are normally equipped with a small sliding or rotating cam lock. The bolt is fastened to the door frame and is engaged by a slot built into the sliding glass door. A flat screwdriver inserted between the door and the frame, and pressure applied either upward or downward, will usually open the latch. Do not strike the screwdriver with a hammer. You may break the glass and damage the frame beyond repair. Sliding glass doors are installed by lifting them into position; they can, therefore, be opened lifting them from the track and removing them. Be very careful as you free the bottom of the glass door to carry it inward, so that, when you lower it, it doesn't fall away from you, thereby breaking the glass and causing possible serious injury.

HOW TO PICK A WARDED LOCK

Warded locks are designed with physical obstructions in the keyhole and in the interior of the lock case. Picking a warded lock is a simple procedure, usually requiring that only one tool be used. The tool must be shaped properly and also be thin enough to avoid the wards when inserted into the keyhole and turned. At the same time, the tool must be strong enough to move the bolt to the unlocked position. This sounds like a rather difficult requirement, but the design principles behind the warded lock are simple (Fig. 10-2). Most warded locks manufactured in this century fit into one of

116

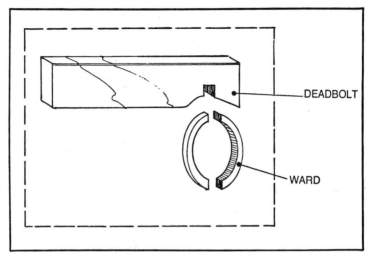

Fig. 10-2. Design principle behind warded lock.

a very few key groups, and can be opened with a pick constructed from an ordinary clothes hanger, called a button hook (Fig. 10-3). To make this pick, take a pair of pliers or Vise Grips and shape a six inch section of hanger to resemble a warded key, including the handle (bow), the blade (stem), and the bit. With some variation in size, the button hook will pick open almost all warded locks encountered.

HOW TO PICK A LEVER LOCK

The picking of a lever lock requires that tension be placed against the deadbolt throughout the course of lifting one or more levers within the lockcase to the required alignment with the post (Fig. 10-4). A lever lock requires the use of an L-shaped tension wrench (constructed of stiff wire) and a lever pick (Fig. 10-5).

Once the tension wrench is in place and exerting pressure to withdraw the deadbolt, the pick is inserted deep into the keyway. Picking begins with the back lever first, raising the lever tumbler gently until you feel a slight give in the deadbolt. This feel may also be transmitted through the tension wrench from the deadbolt, moving just slightly toward the unlocked position as each lever tumbler is picked to the post position. This "give" is a matter of feel, in the form of a sudden release in the amount of pressure required to raise the lever tumbler while applying the turning pressure to the deadbolt. Normally a certain amount of play will develop through use of the lock and through imperfections created by mass produc-

117

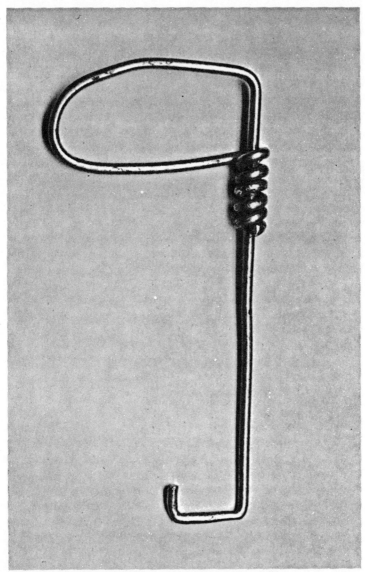

Fig. 10-3. A pick constructed from ordinary clothes hanger.

tion (while manufacturers strive to prevent these imperfections, they do exist). While maintaining turning pressure with the tension wrench, go to the next lever tumbler and repeat the picking process. This procedure must be followed with each lever tumbler in the lock until the deadbolt is retracted into the unlocked position.

As in all picking techniques, one must have proficient knowledge of how a certain lock is constructed in order to understand the messages transmitted by feel to one's fingers. One must know what part is being manipulated at every interval during the picking process. This holds true whether the lock is the simple warded lockset or a more complex type. The basis for this lock knowledge, including construction and operation, is covered in previous chapters. It is important that one have a thorough knowledge of a specific lock type before attempting to pick it.

HOW TO PICK A DISC TUMBLER LOCK

The basic disc tumbler lock described in the preceding chapter is one of the easier locksets to pick, but considerably more difficult to pick than the warded lock. Although there are a number of methods considered standard technique in picking the disc tumbler lock, the simplest calls for the use of a rake pick and tension wrench (Fig. 10-6). The tension wrench is inserted just inside the keyway, stopping short of the first disc. The tension wrench does not reach the full depth of the keyway, as it does in picking lever locks; instead, it is confined to that portion of the keyway which is ahead of the first disc tumbler in the lock. Once the tension wrench is in place, a slight turning pressure (in the direction required to unlock) with the index finger is all that is needed. Too great a turning pressure or tension in picking a disc or pin tumbler lock is a common fault and will prevent the movement of the discs or pins necessary to open the lock. Once the tension wrench is in place, the rake pick is inserted in the keyway and the discs are picked with a raking motion.

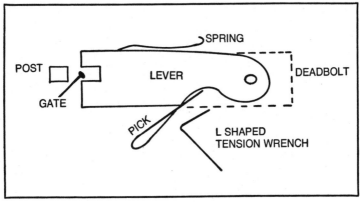

Fig. 10-4. Picking a lever lock.

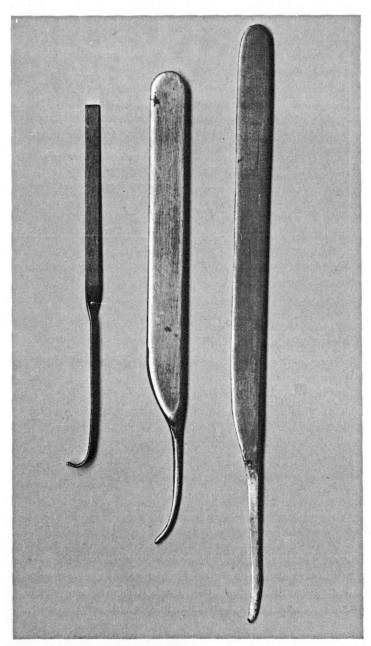

Fig. 10-5. Tension wrenches and picks.

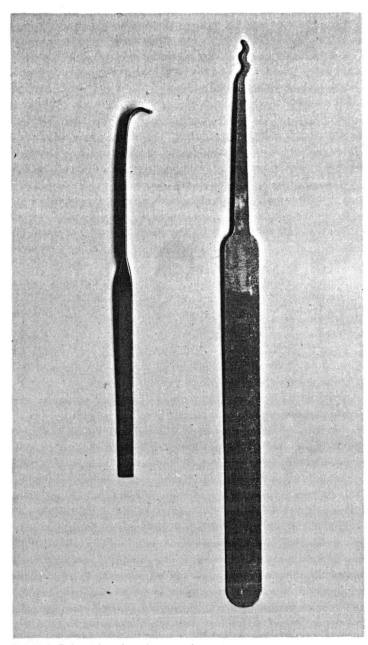

Fig. 10-6. Rake pick and tension wrench.

Several types of picks are used in the picking of disc tumbler locks, but the more commonly used ones are the rake pick and the half diamond (Fig. 10-7).

The half diamond pick, named for its triangular shaped point, is usually thought to be the best choice for picking the disc tumbler lock. This is due to the fact that it may be used to pick the discs in place, as well as raking the disc tumblers in a back and forth motion. This writer has found that the rake pick is much more effective for this purpose and uses the half diamond pick solely for picking the disc tumblers individually when the raking method is ineffectual. But in most instances the raking method is usually effective in picking a disc tumbler lockset.

One must maintain a uniform pressure on the tension wrench while performing the raking or picking of the discs. Again, the feel will be registered in the tension wrench and a slight rotating of the plug as the disc tumbler comes into alignment.

While the raking method is usually the fastest means of picking open a disc tumbler lock, and should be tried first, there are disc locks which will resist this method. When these locks are encountered, revert to the individual tumbler method of picking.

Pick each disc tumbler one by one, until that tumbler is felt to click into proper alignment. This click may be audible or it may be noted only as a sharply increased resistance to further movement of the disc tumbler. However manifested, you will also notice a slight turning of the plug toward the unlocked position. Continue to the next disc tumbler until the lock is finally picked open.

Often a disc tumbler lock may be picked open using a technique known as vibration picking. This is simply utilizing a tension wrench and a snapper pick, which is a pick constructed of spring steel and similar in appearance to a safety pin (Fig. 10-8).

This method of picking consists of maintaining a light turning pressure with the tension wrench while inserting the tip of the snapper pick (vibration pick) into the keyway, just barely touching the disc tumblers. One then uses his thumb, which rests atop the edge of the pick, to depress the top loop of the pick. Let the thumb slide off the compressed part of the pick, permitting it to snap back. In the process, it will strike a light blow against the disc tumblers, causing them to jump up and be held in place by the shear line and the pressure being applied by the tension wrench. Rapid repeating of this snapping of the vibration pick, while maintaining a light turning pressure on the plug with the tension wrench, will usually result in speedy picking of the disc tumbler lockset.

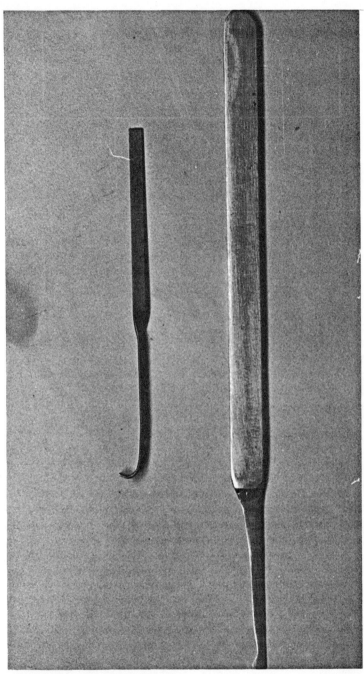

Fig. 10-7. Rake pick and half diamond.

Other vibration picking methods use the Majestic Lockaid picking gun, and the electric vibration pick. All use the same basic theory, which is to give momentum to the disc tumblers sufficient to lift them to, but not beyond, the picked position (flush with the circumference of the plug). To accomplish this, a light touch must be maintained on the tension wrench. There must be some order to the force of the impact used in the vibration method of picking.

With the picking gun and electric vibration pick, control of impact is provided for by an adjustment on the pick itself. With the spring steel vibration pick, it is a matter of feel developed through practice.

As in picking any lock, one must develop patience and skill through practice. The disc tumbler is the type of lock the beginner should select for picking practice. To become skillful at picking either the disc or pin tumbler lock, one must learn the proper handling of the tension wrench. The proper use of the tension wrench is of the utmost importance: if one mishandles it, no amount of picking skill or luck will open the lock.

The natural tendency for the beginner is to apply too much pressure rather than too little. A method to learning proper control of the tension wrench is with the tension wrench in place and lightly held. Observe the fingernail of the index finger; if the nail has a reddish tinge, then there is too much pressure being applied to the tension wrench. Easing the pressure on the wrench will return the nail to its natural color. At the same time, this procedure will provide you with a ready reference for applying turning tension to the lock.

HOW TO PICK THE PIN TUMBLER LOCK

The pin tumbler lock, discussed fully in chapter seven, is the most accepted locking mechanism in use today. It provides a wide degree of application with a reasonably good degree of security, all at a rather modest cost.

While the pin tumbler lock is basically simple in concept, its rather numerous variations and keying possibilities combine to provide a locking mechanism unbeatable to date.

The essential parts of the pin tumbler lock are the cylinder plug, cylinder or shell, the lower pins, the upper pins or driver pins, and the tumbler springs. The plug contains the keyway and the pin chambers (normally from three to seven), which must be drilled in exact alignment, diameter and depth to operate properly with corresponding chambers in the cylinder.

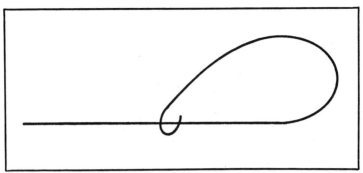

Fig. 10-8. Snapper pick.

The tumbler springs and the upper pins go into the chambers in the shell (cylinder) and the lower pins go into the chambers in the plug. Once assembled and with the proper key inserted in the keyhole, the slightly rounded bottom of the lower pins rest on the bitting cuts of the key blade. Here they raise the bottom of the upper pins and the level top of the lower pins flush with the circumference of the plug, permitting the plug to rotate and unlock the lock. When the key is removed, the upper pins are pushed downward, partially into the chambers of the plug, by the action of the tumbler springs which rest on top of the upper pins (Fig. 10-9). With the upper pins protruding into the plug chambers, the plug will not rotate, and effective locking action is accomplished.

Since the pin tumbler locking mechanism is the most popular lock in use today, and because it offers greater resistance to picking than the disc tumbler or warded locks, numerous techniques for picking have been devised to attempt to defeat these locks.

The first method of picking is the feel method which employs the same basic principles as in picking the disc tumbler lock. After determining the proper rotation of the plug to unlock the locking mechanism, select a tension wrench and pick for picking the lock. This selection will depend considerably on the size and configuration of the keyway.

The tension wrench is placed just inside the keyway, stopping short of the first pin tumbler. A light to moderate tension (pressure) is used and the pins are picked one at a time, beginning with the pin which is bound (offers resistance) the tightest. Care must be taken not to lift the lower pin past the circumference of the plug and into the upper pin chamber. Bringing it down again may require releasing the pressure entirely and starting over. The method used to avoid over-lifting is to apply just enough force with the pick to lift the pin up against the downward spring pressure involved. Being extremely careful not to overcome the added resistance felt when the lower pin contacts the shell at the shear line.

In the mass production of pin tumbler locks, it is a near impossibility for the manufacturer to drill all the pin chambers in every plug in perfect alignment. This imperfection makes it possible to raise one pin to the shear line at a time. Upon reaching the shear line, the plug rotates slightly toward the unlocked position. The next one raises the next binding pin until it also reaches the shear line. Again there will be a slight turn of the plug.

The turning of the plug toward the unlocked position as each pin reaches the shear line is very slight. Practice to develop the proper feel for it is important. Continue to pick the pins in the order of the next bound pin until the plug completes its turn and the lock unlocks.

The next method of picking is known as the raking method which was discussed as a means of picking the disc tumbler lock. While it is not as sure as the feel method in picking open the pin tumbler lock, it could be employed as a short cut. If one is fortunate to align all the pins at the shear line, he doesn't have to bother searching for each bound pin and moving it individually to the shear line.

The theory is to move the rake pick quickly back and forth under the lower pins, while applying light turning pressure with the tension wrench. This action causes the pins to bounce up to the shear line and dangle there as the plug is turned further out of alignment with the upper pin chambers. The plug eventually turns to the unlocked position.

Vibration picking may be used, utilizing the spring steel snapper pick, the Lockaid pick gun or the electric vibrator pick. In all the vibration pick methods, a light turning pressure is applied with a tension wrench.

The function of the vibration pick is to strike all the lower pins at once, bouncing the upper pins into the upper chambers while the lower pins remain in the lower chambers. When this occurs, no pins obstruct the shear line and the plug is free to turn, unlocking the lock.

On occasion, a pin tumbler lock may sometimes be opened using a technique called *rapping* (sometimes called *cracking,* a form of vibration picking). A tension wrench is inserted in the keyway and light to moderate tension is applied. At the same time, the face of the cylinder is struck sharply with a plastic mallet or hammer

126

Fig. 10-9. Pin tumbler with key removed.

handle. This procedure uses the law of inertia, in that the rapping forces the springs and pins to gravitate toward the force of the blows. Care must be taken in doing this, however, to avoid damaging the lock. Approach this method with caution.

When for some reason it is impossible (or, due to damage to the keyway, (not practical) to pick a lock, a forcing method can be used. A large pair of vise-grip pliers can be tightened over the outer rim of the cylinder and then turned, forcing the set screws loose and defeating the lock.

HOW TO DRILL A PIN TUMBLER LOCK

Sometimes a malfunction in the lock will prevent it from being picked. Drilling a pin tumbler lock is a less destructive means of opening the lock without the key than forcing the lock cylinder. Almost any type of drill may be used, including a hand operated drill,

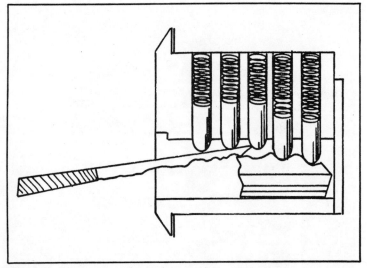

Fig. 10-10. Extractor inserted all the way into keyway.

but the small one-quarter-inch electric drill is faster. The bit size is not of great importance, but the small bit will make the task easier when using a hand drill. Normally a 1/16 to 1/32-inch twist steel drill bit will do the job.

Make the hole just above the lower tumbler pins near the shear line. Once the hole is drilled to approximately one and a half inches, take a tension wrench and turn the plug toward the unlocked position. If the plug refuses to turn, take a thin stiff section of wire and push the upper pins above the shear line to rotate the lock open.

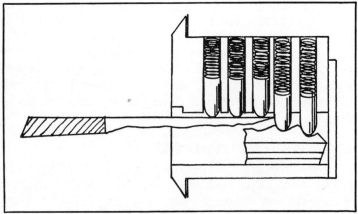

Fig. 10-11. Lifting up forward pins with stem of extractor.

HOW TO EXTRACT A BROKEN KEY

A piece of broken key can be removed with a length of fret-saw blade, coping-saw blade or a straightened fish hook. First, check the plug and make certain it is lined up with the upper pin chambers. Unless the plug chambers are aligned with the upper pin chambers, the broken key will not budge because the pins are not free to move into the upper chambers as the key is withdrawn.

Insert the extractor all the way into the keyway (Fig. 10-10) until you can feel the hook grasp the first cut in the key blade. Next, gently lift the forward pins up with the stem of the extractor (Fig. 10-11) by raising the extractor until it is parallel with the top of the keyway. Once it is in this position, slowly but firmly withdraw the extractor and the key will come out.

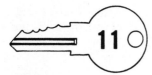

Safes

Many centuries ago the ancient Chinese developed a ring puzzle toy known as a *letter lock*. It consisted of a number of rings, also called *tumblers*, inscribed with letters or numerals and threaded on a spindle. When the rings were rotated so a certain word or series of numbers appeared, the spindle could be withdrawn, because slots cut into the rings fell into alignment.

This ancient letter lock was the forerunner of the simple number combination lock (commonly called the *keyless lock*) used in England at the beginning of the seventeenth century. At first these locks were used only as padlocks and on strongboxes. Today this same concept is found, to secure bicycles and motorcycles, in the form of a *combination chain lock*. This same basic principle is also the basis of the extremely efficient combination locks used to protect home and office safes as well as most vaults.

During the last half of the nineteenth century, the need developed for safes and strongrooms. The combination lock proved to be the most secure form of lock. The number of possible letter or number combinations was seemingly endless.

It can be said that the modern combination lock is an American invention (refined by such notable lock pioneers as Yale and Sargent) improving upon the ancient Chinese craftsman's letter lock.

If any lock can be said to be unpickable, it would have to be the modern combination lock. Except for a limited number of experts in the lock trade, most others would fail in their attempts to open a modern combination lock through manipulation.

Therefore, the only alternative is the route taken by most locksmiths (as well as thieves); that is, cut the safe open with carbide-tipped drills or torches, or to blow it open with explosives.

A modern combination lock with three rings (tumblers) and one hundred numerals on the dial presents one million possible combinations. In addition, this combination lock may be X-ray proof, consist of anti-manipulation devices and contain relocking mechanisms.

The smaller home and office safes first appeared around 1850. Because safes are normally constructed of iron and steel and are not subjected to a great deal of wear, they will last forever—or until they are attacked by some form of safecracking technique. These techniques could include cutting, ripping, peeling, prying, punching, drilling, burning bars, x-ray, or other form of manipulation. In spite of the variety of safecracking methods, many safes constructed over one hundred years ago can still be found in use today.

While there are numerous combination lock designs, fundamentally all consist of a single spindle affixed with an outer numeral dial, to which are placed from three to four circular discs (called tumblers or rings) mounted to the rear of the spindle.

PROPER OPERATION OF A COMBINATION LOCK

In order to properly work a three-tumbler combination lock, one must first clear the tumblers. To do this, you spin the exterior numeral dial a minimum of four turns (complete rotations) to the left. This permits the spindle to pick up all three tumblers and after a minimum of four turns (it can be more), but not less than four turns, you then stop at the first number of the combination. Using a standard factory set combination of 50-25-50 with zero (0) being the opening number, you would proceed in the following manner:

direction	turns	combination numbers
Left	4	50
Right	3	25
Left	2	50
Right	1	0

Spin the numeral dial four times to the left and stop at 50 on the opening index mark above the dial. The dial is then rotated to the right three complete turns, stopping at the second number, 25, on the third turn. Next, spin the dial to the left, making two complete cycles and stopping at 50, the third number on the opening index mark. Finally, make one turn to the right to 0 (commonly called the opening reference) and, depending on the type and model of safe, either open the safe by turning the safe handle or continue to rotate the dial past 0 until this action retracts the locking bolt.

Although this may seem like a complicated and laborious method of operating a lock, once a person becomes accustomed to the procedure it can be quickly performed.

What appears to be a complicated alternating reversal of the disc tumbler rotation is necessary to properly operate the inner tumblers so that when these tumblers, with their individual gates (openings cut into the tumblers), are brought into alignment with the bolt retractor, it drops into the gates. This allows the locking bolt to be withdrawn, and the safe door is then capable of being opened. This is the same principle used in lining up the notches in the rings of the ancient Chinese letter locks, but with modern day refinements.

Always dial numbers precisely: do not turn back to regain a number once you have passed it. You must redial the entire combination, starting with step one; rotate the dial a minimum of four times past the first number of the combination.

CHANGING A KEY CHANGE COMBINATION LOCK

To change the combination, first dial the present combination on the changing index (Fig. 11-1), stopping on the third number. Again, let's use the factory combination of 50-25-50: rotate the dial a minimum of four times to the left, stopping with 50 at the changing index. Turn the dial to the right three complete turns, stopping with 25 at the changing index on the third turn. Rotate the dial to the left two turns, stopping with 50 at the changing index on the second turn. *Do not go to 0.* Dial the three digits 50-25-50 only. This properly aligns the discs with the keyway to the rear of the lock, permitting you to insert the change key.

Next, take the change key furnished with the safe and insert it into the keyway at the rear of the lock case. After checking that the key is firmly seated into the lock, turn the key clockwise one quarter turn. Then spin the dial at least four times to the left. This action clears the old combination off the discs. You are now ready to set a new combination.

In selecting a series of numbers for your new combination, avoid exclusive use of fives and tens. Distribute combination numbers over the entire dial, i.e., 7-42-93. Do not use special occasions such as anniversaries, birthdays or any other combination of numbers that a would-be thief might be able to acquire to defeat the lock.

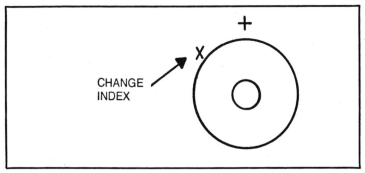

Fig. 11-1. Dialing present combination of changing index.

In setting the new combination, in the same format used to remove the previous one. Remember to use the changing index. Also, keep in mind that each time you pass the desired number, that's considered one complete turn.

For the purpose of this exercise, let's use 50-25-50 as the new combination.

To set the new combination, turn the dial to the left (counterclockwise), stopping when the number 50 is aligned with the changing index on the fourth turn. Next, turn the dial right (clockwise), stopping when 25 is aligned with the changing index on the third turn. Then turn the dial left (counterclockwise), stopping when 50 is aligned with the change index on the second turn. After ensuring that the dial is reflecting 50 on the change index, turn the change key counterclockwise one quarter turn and remove.

With the safe open, dial the new combination 50-25-50, including the opening reference 0, this time using the opening index. Test the new combination with the safe open several times before closing and locking the safe.

CHANGING A HAND CHANGE COMBINATION LOCK

To change the combination on a hand change safe, first open the safe and remove the backplate to the rear of the safe door. Next release the locking bars, locking the safe open. Spin the dial to the left a minimum of four turns, stopping at 50 on the opening index. Now remove the two screws securing the back of the lock case.

Remove the lock ring and take the disc tumblers off one at a time. You will note that each tumbler is numbered one through three. Tumbler number one is the first number of the combination and is the first tumbler to be reassembled; then two; and finally tumbler number three.

To change the number for each tumbler, using 50-25-50, take tumbler number one and press the middle section out. Next align the reference mark with the number 50 on the tumbler ring and press together. Then go to tumbler number two and align the reference mark with 25 on the tumbler ring. Next align 50 with the reference mark on the third tumbler. Reassemble the tumblers, beginning with tumbler one, and continue through three. Replace the lock ring and place the back plate against the lock case. Tighten with the screws.

Finally, try the new combination and bolt linkage several times before closing and locking the safe.

MAINTAINING SAFE SECURITY

There are many definite steps which can be taken by the safe owner to reduce the constant vulnerability to unauthorized disclosure of the full safe combination, partial combination, or inner construction of the safe. Always keep the safe door closed, especially if it is in the range of observation. A professional can determine the type and construction of the container by seeing the interior, back of safe door, number of locking bars, and so forth.

After opening the container, always move the dial away from the last number of the combination *immediately*. When locking the safe after use, spin the dial at least four complete turns to the left *and* the right to ensure scrambling of the combination.

Safe combinations should be changed every six months, as well as simultaneously with the departure of anyone entrusted with the combination or whenever you suspect the combination may have been compromised.

There are on the market today many models of safes adequate for storage of important papers such as deeds, wills, and insurance papers. These containers offer a small degree of security and are normally fire-resistant. But if you are in the market for a burglar-resistant safe, always check for the Underwriters' Laboratory label which states that the container meets the minimum requirements: "Torch and Tool Resistive or Torch, Tool and Explosive Resistive".

Burglar-resistant safes approved by the Underwriters' Laboratory must be constructed of tempered steel at least one inch thick and have doors made of steel at least one and one half inches thick. The combination lock must be UL listed and must be equipped with a relocking device. The doors must be constructed of special metal alloys resistant to carbide drills, and the safe body

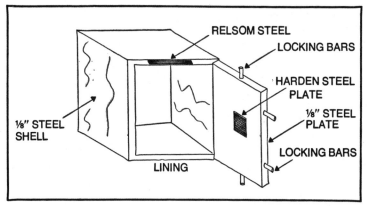

Fig. 11-2. Burglar-resistant safe construction of Relsom, a new metal bonded with copper plate which is torchproof, drillproof and shatterproof.

must be cast or welded plate. If the safe weighs less than 750 pounds, it should be equipped with anchor bolts to secure it in place.

To safeguard important legal papers and priceless objects such as coin or stamp collections, the only past adequate protection has been to rent a safe-deposit box. But most safe-deposit boxes are too small for bulky valuables and too inconvenient for any object or papers that would be used frequently—silver flatware, collections, cameras, financial records and the like. The best way to protect such items is to store them in a burglar and fire-resistant safe (Figs. 11-2 and 11-3) or vault room (Fig. 11-4).

When choosing a safe, determine first whether you are more concerned about burglary or fire. Protection from both is rarely combined in a single safe, and when it is, it is quite expensive.

Burglar-resistant safes are made of thick tempered steel castings, and are equipped with sophisticated combination locks and hardened steel bolts. But papers inside may char or burn in a fire. A fire-resistant safe, while cheaper, offers little or no burglary protection. It is usually constructed of double layered sheet metal and filled with several inches of fire insulating material. Even if equipped with a combination lock, a determined burglar can open a fire-resistant safe in a matter of minutes by peeling back the sheet metal skin.

One variety of fire-resistant safes does more than just block the flow of heat. Its insulation is a crumbly, crystaline mixture of lightweight concrete and granules of mica containing as much as 25 quarts of water. Ordinarily the water is locked into the crystal structure. But a fire disrupts the crystals, releasing the water and

135

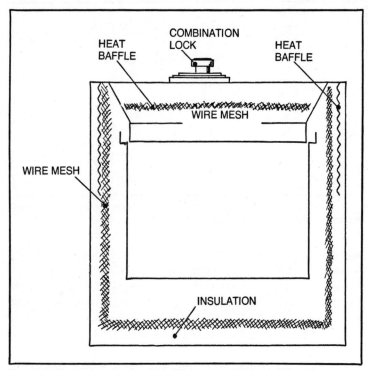

Fig. 11-3. Fire resistant safe with baffles to prevent heat buildup.

turning it into steam. In the process, the water absorbs great amounts of heat. The steam filters through small openings inside and outside the safe and may crumple and discolor papers, but they will remain legible and flexible enough to handle. Since the water of crystalization is so easily released by heat, old safes may have lost their protective capability; a used safe should not be purchased.

Always look for the Underwriters' Laboratories (UL) label on any fire-resistant safe you purchase. The label will state the type of safe, what tests it has passed and the insulating capability. The label will also specify the number of hours the safe will withstand a fire. A one hour safe is adequate for most home uses, but two and four hour safes are available and should be considered for business use. The label will also indicate the maximum temperature inside the safe during the course of the fire. Example: Class 350F for a safe designed to protect papers; Class 150F for a safe designed to protect magnetic tape or transparencies.

Because the water in the insulation of a fire safe is so easily released, it can make the interior damp. To minimize this problem,

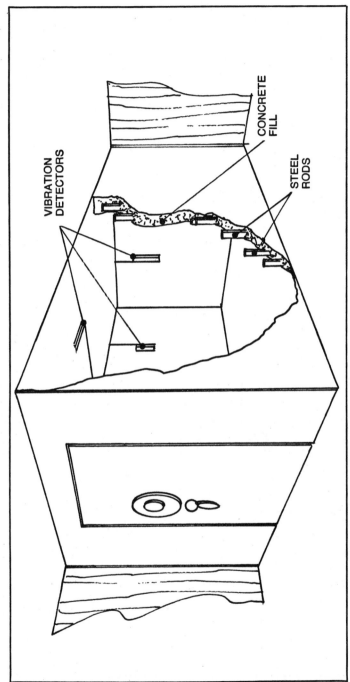

Fig. 11-4. Vault equipped with steel rod reinforced concrete walls, 6 to 12 inches thick.

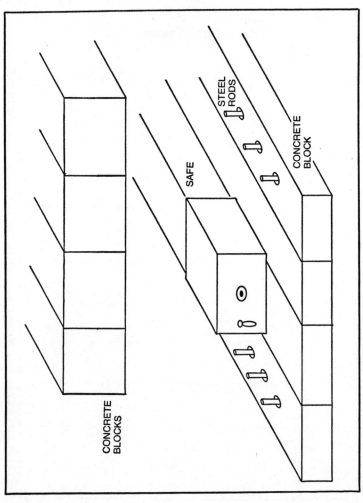

Fig. 11-5. Installing a safe in a concrete anchor vault, constructed by placing the safe within concrete and steel reinforced blocks.

138

locate the safe away from direct sunlight, radiators, and other sources of heat, and open it two or three times a week to air the interior. If dampness occurs, leave a small bag of silica gel in the safe to absorb moisture. Use airtight plastic bags to seal documents or objects especially susceptible to water damage.

If you should have a fire, locate your safe as soon as you can safely enter the building. Have the firemen wet it with a fine hose spray until it is cool enough to touch. Open the door immediately, while someone stands by with a fire extinguisher; the contents may burst into flame when the steam which has been covering them is released. Last, replace any safe that has been through a fire: its fire-resistant capabilities have been destroyed.

Where a fire-resistant safe can provide absolute protection for a specified period, a burglar-resistant safe cannot offer that guarantee. No safe can totally ward off today's sophisticated cutting tools. The majority of home burglar-resistant safes can withstand sledge hammer, drills and prybars for only a few minutes. UL listed burglar-resistant safes, which give far better protection, are designed primarily for business use and are expensive. Whether buying a safe for home or business, evaluate what you need to protect and buy accordingly. If you can afford a burglar-resistant UL listed safe—buy it. If not, check the safe you intend to purchase, part by part. It should, at the minimum, have a UL approved three-position combination lock. It should also have a UL listed relocking device to fasten the locking bolts in place if a burglar attempts to drill through or punch the lock, and a hardened steel plate to protect the lock housing.

The type of safe you buy and the method of installation depends a great deal on the construction of your home or business, the amount of storage space needed and the type of protection you desire. If you desire, you can build a concrete anchor vault around a small floor safe (Fig. 11-5). Most safes are equipped with anchor bolts or mounting bolts.

Automotive Locking Systems

Automobile locks vary considerably in the types of locking mechanisms and lock arrangements. There are many other methods of gaining access to an automobile which are quite often quicker and easier than attempting to pick the lock.

Probably the most common point of entry is the wing or vent window, or, in the newer models, the rubber molding. Anyone intending to engage in automobile lock work (or desiring to save approximately fifty dollars or more in service calls) will need some special tools. Both the curve and length of tools are important to their overall effectiveness.

The tool being used should first be lubricated with a glycerine-based hand lotion or hand cleaner before inserting it through the weatherstripping. This not only makes the tools work easier, but helps prevent scarring, tearing or any other damage to the weatherstripping.

To open most front wing windows, take the lever latch tool and insert it through the weatherstripping between the wing window and the window trim. Manipulate the tool by controlling the depth of penetration along the curve. At the same time use a rocking action to move the window lock into the unlocked position.

One other type of wing window lock has a lever latch equipped with a plunger at the pivot of the latch. The plunger deadlocks the latch against rotation, unless the plunger is first pushed in and held until the initial stage of rotation has been accomplished. This re-

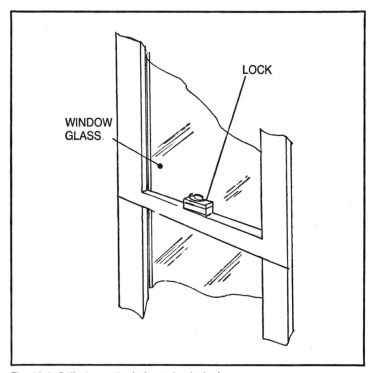

Fig. 12-1. Stiff wire method of entering locked car.

quires that another tool be inserted through the weatherstripping and the door window to depress the plunger and hold it in that position while the other tool is used to rotate the latch. Normally, the tool used for depressing the deadlocking plunger has a slight curve at the tip. The tool's only purpose is to depress the plunger.

Another means of access, when the door window is fully raised and the door is locked, consists of inserting a stiff bent wire (Fig. 12-1), and using it directly on the locking mechanism. Tripping of the lock mechanism may often be done by pulling up on the wire once the tip of the wire has been positioned under the lock linkage. At other times, the bent tip of the wire must be pulled up against the linkage and then rotated to trip the locking mechanism. With practice one gets the feel of what is required to open the door.

Automobiles using a rocker type of locking mechanism may be tripped by using a thin piece of flat spring steel stock. First, coat the tool with a glycerine-based hand lotion or hand cleaner to protect the trim and rubber molding. Insert the tool next to the glass or between the weatherstripping/molding and the metal of the door.

Feel for the lock linkage with the notched end of the tool, then move the linkage up and down until the lock moves into the unlocked position.

Another avenue of access to most automobiles is through the firewall, reaching the door locks with a long stiff wire to push the lock button into the unlocked position.

Since automobile door locks on most newer models are normally held in place by a retainer clip, as a last resort the lock can be punched out. First, insert a bent piece of wire into the keyway and bind it to prevent the lock mechanism from falling into the door frame. Once out, the automobile can be opened by pushing up on the linkage attached to the lock mechanism.

Security Alarm Systems

From biblical times, numerous forms of alarm systems have been created in the hope of detecting intruders. Early alarms were rigged out of cord attached to a noisemaker (called a trip wire). This principle is still the basis of today's modern alarm systems. A detector senses an intrusion, passes an alarm signal to the control, and an alarm is sounded.

The era of modern security systems began over 100 years ago with the invention of electrically-operated alarms. Electricity made it possible to protect many different locations by connecting all of them to a single alarm annunciator. Electrically-operated alarms provided increased reliability in the detection mechanism as well as in the alarm signal itself.

A major breakthrough in the central station alarm concept came in 1859 with the transmission of an electrical signal from one building to another via telegraph lines. The first security systems installed used the Western Union telegraph lines. As other alarm companies started up, they too relied on Western Union lines whenever they were available. Where there were no telegraph lines, alarm companies strung their own wire from the protected premises to the central station, where the alarm was received.

The rapid expansion and growth of telephone systems provided a ready wiring network for security system companies. Security alarm companies could now extend their service by using telephone company lines to connect to more points, both local and remote. And they could do this at an economical price.

With the advent of the electronic microchip, today's security systems offer numerous advances in alarm technology, such as telephone dialers that alert police through pulse dialing; hi/low sirens; systems that turn on spot lights; radio frequency (RF) transmitters and receivers. With the use of electronics, systems are becoming more reliable and costs are coming down.

CHOOSING A SECURITY ALARM SYSTEM

Until recently, security alarm systems simply weren't considered necessary in the home or in most small businesses. Because the demand was so small, manufacturers did not attempt to develop and produce economically practical or mechanically simple security systems for the homeowner or small businessman. The individual who, for one reason or another, felt that he needed the protection of an alarm system, had to settle for the extremely expensive commercial systems or design and build his own.

However, the spiraling crime rate over the past ten years (and increased burglaries in particular), has created a great deal of activity in the security alarm field. There are presently many low-cost alarm systems which an individual can install with little difficulty. Also, instructions and diagrams have been published in a number of monthly magazines which will allow anyone so inclined to design and build a security system suited to his needs.

Of course, the larger, more complex, and more costly commercial security alarm systems are better and more available than ever before, and the individual willing to pay the cost can make his selection from a wide variety of makes and models. He can, if he chooses, make his home or office so sensitive that a mouse running across the floor will trigger an alarm.

The amount spent on a security alarm system depends on two factors. One factor is the type of place needing protection, and its location. If it is located in a high crime area or would attract burglars due to the types of merchandise or money stored, it should have an extensive alarm system.

The second factor is whether the individual is willing or able to assemble and install the security system, thereby reducing the cost substantially.

Small business and residential systems, complete and installed, cost from $650 to $2500, with the average at approximately $1500. Installations in vulnerable areas or in locations with many entrances and windows, require numerous magnetic contacts, vibration contacts, and window coverage. This can push the total cost well over $2000. Over half that cost is labor for installation. In contract, the individual who installs his own system would pay only for the components, which would run approximately $300 for an open circuit system, $250 for a closed circuit system, and around $500 for a combination fire/burglar system.

Some of the material required for a closed circuit system are:

☐ Control panel (with or without delay entry/exit)
☐ Alarm bell or siren
☐ Magnetic contacts
☐ Door shunt switch
☐ Vibration switches
☐ Tamper switches
☐ 6 Volt batteries or transformer

It is possible to start your security alarm system small, covering just the primary doors and really vulnerable windows, expanding it whenever you wish. But first you must make a definite decision on the basic type of security system to be installed: will it be an open circuit or closed circuit system? The reason for this is that many of the components, particularly the magnetic contacts, are not interchangeable. Some points to consider in making your choice are:

☐ Least expensive and most dependable.
☐ Easiest to install and troubleshoot.
☐ Simplest to operate.

The Open Circuit System

The above remarks describe the *open circuit* system. An open circuit system has magnetic detectors or contacts of the "normally closed" type. That is, their contacts are separated when the doors or windows are in the normally closed positions. But opening the door or window releases contacts, causing them to close. This allows current to flow through the wires to sound the alarm. All contacts and detectors are wired in parallel: current flows only when any contact/detector switch makes contact.

A disadvantage of this system is that it becomes inoperative if a transmission wire is cut, a contact or terminal wire becomes loose, or some similar condition occurs. However, you can conceal the circuit wiring against tampering.

The vulnerability of the open circuit system is minimized by a test switch or key position which sends current through the main circuit wiring and reveals any line breaks. This test lights a small warning light or sounds a buzzer, but bypasses the alarm bell which does not ring during the test. The above test does *not* show whether each individual detector is in working order.

All it takes to set the open circuit alarm system is to turn the key switch to the "on" position. The key switch can be placed in any convenient room or it can be panel mounted (the key switch is mounted in tamperproof control panel). False alarms will immediately occur if doors or windows are left open when the alarm switch is turned to "on" or when it is in the "on" position. You can reduce this possibility by having bypass switches located near frequently-used doors or windows. However, if located too close to a door or window, these switches become a weakness in the system: someone breaking a door or window pane can reach through and push the switch, thereby allowing free access through that door or window.

The Closed Circuit System

In the *closed circuit* security system, low amperage current continuously flows from the battery or recharging pack, through the detector switches, to the supervising relay within the control panel. The detector switches are the normally open type (Fig. 13-1). The contacts are in the closed position when the protected area (windows/doors) are shut against the magnetic contacts. Current from 3 to 12 volts flows continuously. If for any reason the current is interrupted, even momentarily, the supervising relay opens and releases the contact. This sets off the alarm bell, siren or telephone dialer. The alarm will continue to be activated until the control panel key switch is turned off and the drop relay reset.

Current to the relay is normally interrupted when a door or window is opened or a similar break occurs in the protected circuit. These breaks could take the form of rattling of windows protected with vibration contacts, breaks in the protective circuit wiring, or erratic operation of the supervising relay. With today's new solid state equipment, false alarms caused by anything other than breaks in the circuit are rare. But when such a condition occurs, the alarm cannot be reset until the fault has been located and corrective action taken. Dividing the alarm system into zones can help to pinpoint the location of difficulty. It also allows setting the alarm for functioning areas during zone disruption.

The closed circuit system requires more sophisticated equipment and the circuit installation must be precisely wired. One chief problem is the tendency for false alarms. However, a distinct

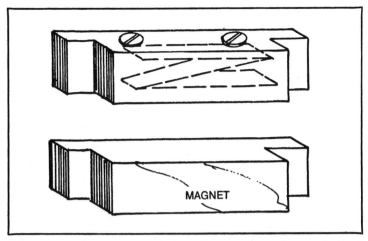

Fig. 13-1. Normally open type of detector switches; copper contacts are in closed position when magnet is shut against upper portion.

advantage is that, unlike the open circuit system, if there is a problem the system will not set up until it is corrected.

Normally Open and Normally Closed Contact Switches. The term *normally open* and *normally closed* detector switches can be confusing to all but the experienced installers. The *normal* position of a contact switch is that condition it is in when it is in the rest position. That is, unenergized (no current flow), or uncompressed (no pressure). Thus, a button switch that is normally open (N.O.) will be one in which the contacts are open (separated) when the button extends from the casing. When it is pressed into the casing as in closing a door or window, the contacts close and the circuit is complete. A "closed circuit" system then requires "normally open" contact switches.

For the open circuit system, you would install "normally closed" (N.C.) contact switches. When installed, if the button is depressed by closing a door or window, the contacts open and no current flows (Fig. 13-2).

When buying security alarm system components, it is a good idea to be clear about the type of system you are planning to install. Some contacts are clearly marked N.O. or N.C., but many others are not or are marked on the package. Normally open and normally closed contacts are not interchangeable.

The thing to always remember is to select the opposite of the security system you are installing—"normally open" for closed circuit systems and "normally closed" for open circuit systems.

147

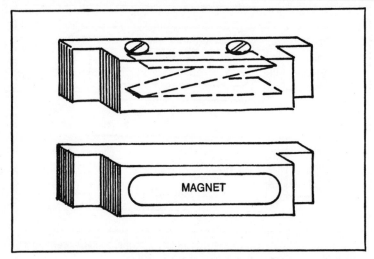

Fig. 13-2. Normally closed contacts; copper contacts are in the open position when magnet is shut against the upper portion.

Security Alarm System Power Sources. The current for most security systems is provided by battery, transformer or a combination of the two called a recharging pack (Fig. 13-3). The recharging pack is a complete power supply providing 6-12 volts of power to the protective circuit, and has the ability to operate up to two alarm bells or sirens and provide six volts for a telephone dialer or transmitter hookup. It is equipped with sealed nickel cadium (NICAD) rechargeable batteries, providing standby power in case of electrical power failure.

Some supervised closed circuit systems use 3-6 volt batteries for the relay and the annunciator, and/or 110 volt house current for the annunciator with 6-12 volt battery backup. The recharging pack described above eliminates all of this and provides a trouble-free, totally reliable power supply system (the ultimate power system). It allows the alarm system to function independently of the home electrical system during power failures and then recharge via a trickle charge when power resumes.

PURCHASING SECURITY ALARM SYSTEM EQUIPMENT

Substantial savings can be realized by first planning what alarm equipment you will need, and then ordering all components at once. Unit charges for single item purchases are often considerably higher than volume purchase. Often, suppliers will discount a total alarm system package by 10 to 40 percent under retail. Also, suppliers will usually pay shipping on orders totaling a certain minimum dollar amount. It will be well worth your time and money to comparison shop suppliers.

The quantities of components needed will depend on the scope and extent of the security system you wish to install. The particular type of equipment will also depend on the individual situations in your home or business—kinds of doors and windows, degree of vulnerability, and many other factors. An extensive line of equipment has been developed to cope with any situation.

The sentinels to all security alarm systems are the detectors, devices that stand guard over all accessways into the house or business and trigger an alarm when intrusion is either threatened or accomplished. They are explained here to allow you to select that best suited for your needs.

Magnetic Detectors. By far the most effective, dependable and versatile components, the magnetic detectors consist of two parts, each completely encased in dustproof/weatherproof plastic. One part contains the complete switch (Fig. 13-4), with two screw terminals and contact points. The other part is the magnet (Fig. 13-5), which is attached to the moving portion of the installation (the door or window sash). The two parts are installed close together. When the sash is raised or the door opened, the magnetic contact moves away and shifts the internal points, causing an alarm condition. These contacts lend themselves to numerous uses and cost approximately $3.00 to $4.00 a pair.

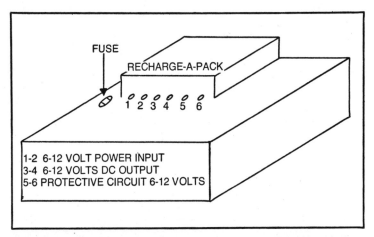

Fig. 13-3. Recharging pack provides constant 6-12 volt power for the total alarm system and emergency nicad battery backup power during electrical interruption.

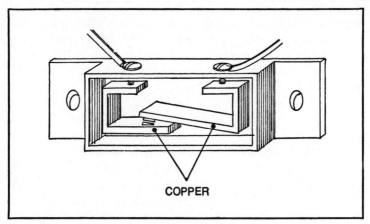

Fig. 13-4. Magnetic detector with complete switch.

Tamper Switch/Plunger Contact. Another detector lending itself to many uses is the tamper switch. It may be used on casement windows, alarm bell/siren boxes or control panels. The switch is constructed of copper points encased in plastic (Fig. 13-6) and costs less than $2.00.

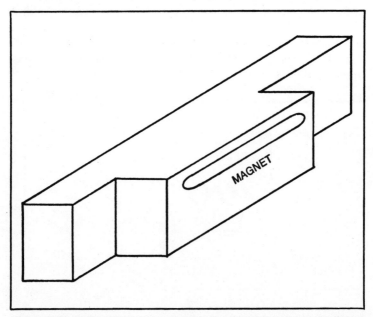

Fig. 13-5. Magnetic detector with magnet, causing copper plates or points to separate or come together depending on whether it is an open or closed circuit system.

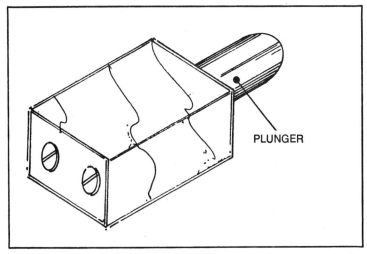

Fig. 13-6. Tamper switch, available in both open or closed mode, lends itself to many adaptations.

All-Purpose (Bullet) Detector. This is a beveled button used primarily on doors, but it is also suitable for double-hung windows. The button is installed in the hinged side of the door frame, recessed into the frame. When the door is closed, the button is depressed. These contacts sell for approximately $1.25 each.

Entrance Shunt Lock. Entering or leaving the residence or business while the security system is armed is accomplished via the shunt lock. Installed near the entrance, it either turns off the alarm or bypasses that particular entrance. The shunt key lock is available in double bit or ace (round) key (Fig. 13-7).

There are numerous other types of detectors available for security systems on the market. The extra effort expended to add

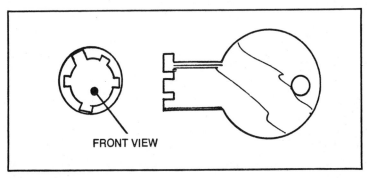

Fig. 13-7. Ace key.

one or more of these detectors can go a long way toward defeating the resourceful burglar who studies and defeats the common perimeter protection system. Some of these detectors and their applications are:

Floor Mats. Pressure-sensitive mats wired with open or closed circuits to make or break contact when stepped upon are used as backup to perimeter security systems such as patio sliding doors. The mats can be placed under regular carpet or loose rugs at any location.

Door and Window Traps. Pull trap wire contacts with alarm switches can be used in even the seemingly most hopeless situation, such as windows and storage areas. If entry is attempted, the trap wire is pulled out of its spring-loaded retainer. An alarm is triggered. The traps can be placed either horizontally or vertically across the area (Fig. 13-8).

For open circuit systems, an insulated trap is used with a insulated clip. Security system wires are connected to both sides of the trap. When the clip is pulled out, the trap contacts meet, sounding an alarm.

For closed circuit systems, one wire is connected to the top bracket and the other wire is connected to the trap wire, maintaining constant current in the system. Pulling on the trap will break the circuit and sound the alarm.

Panic/Emergency Button. This permits sounding the alarm, either bell/siren or telephone dialer, in the event of an emergency. Buttons can be installed near the front door, master bedroom, or hallway in a home. In a business, the button can be installed as a holdup button, silently summoning the police through the use of a telephone dialer or transmitter.

Automatic Telephone Dialer. This automatically calls any pre-programmed (recorded) telephone number such as the police department, fire department, emergency services or individual home and relays a recorded message to summon help. The electronic pulse dialer, which has been approved by the Federal Communications Commission (FCC) and the telephone company, will not interfere with normal phone usage and can be programmed to send up to five messages per channel with two channel capability. Unless used on a private leased line (which can be expensive), the dialer may run into with such difficulty as a busy number.

The reliability of the dialer can be seriously compromised if the exterior telephone wires are not secured with a metal conduit or hidden to prevent someone from clipping the wires. All dialers should be backed up by an annunciator such as a bell or siren.

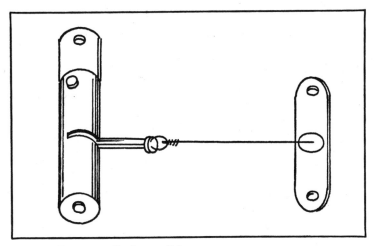

Fig. 13-8. Trap wire with alarm switch.

Photoelectric Systems. Photoelectric systems transmit invisible pulse modulated beams from projector/transmitter to receiver. Interruption of the beam sets off the alarm. The basic system is designed for interior use, but through military application, systems have been designed which will function outside even in rain and dense fog. Protection may be extended over a large area through the use of reflective mirrors.

Fire Detectors. Fire protection is, of course, essential for every home or business. With today's double digit inflation, equity in one's home or business becomes even more necessary. Smoke detectors, fixed temperature detectors and warning horns should be in every home or business. These systems can be installed separately or as part of your security system, providing overall security.

Installing the Security Alarm System

The first and most important step to installing a security alarm system is to design a layout or diagram. You can save time, as well as provide a professionally finished product, if you take the time beforehand to complete a sketch or diagram of what will be protected, where the control panel will be, and what equipment will be needed.

WIRING

All necessary wiring for the system's circuitry starts at the control panel. Separate wires to each detector will result in unnecessary wiring. Possible, professional installers link most detectors in a single run of wire called a *loop.* Whenever careful planning will make this possible. Annunciator lines, transformers and panic switches are carried on separate lines.

CONTROL PANELS

Modern solid state panels greatly simplify alarm installations and improve the efficiency and dependability of the system. Today's control panel assures proper hookup of the critical elements of the system and also provides for additional aspects to make the overall system more functional, such as test indicators, fire lines, panic lines, zone arrangements and on/off signal lights. These control panels are available for approximately $55 and up. They are extremely reliable.

CIRCUIT WIRING

Open circuit (O.C.) and closed circuit (C.C.) systems are wired differently. In the open circuit system, the normally closed switches are wired in parallel; that is, both line wires are connected to the contact switch—one wire on each terminal. The circuit wire then continues on and is connected in the same manner (Fig. 14-1). There is no reason to follow polarity for an open circuit system. The

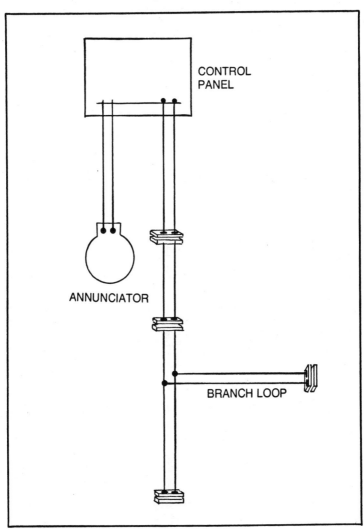

Fig. 14-1. The open circuit alarm system consisting of normally closed detectors wired in parallel.

wire continues to the last contact where it simply ends. No return is needed, as the two-conductor wire serves as the complete circuit. Whenever the contacts close on any of the detectors, forming a complete circuit, current flows back to the control panel and sounds an alarm.

Normally, in closed circuit systems, open detectors are wired in series. A pair of wires starts at the control panel and completes a loop back to the panel. Only one of the wires is cut and attached to

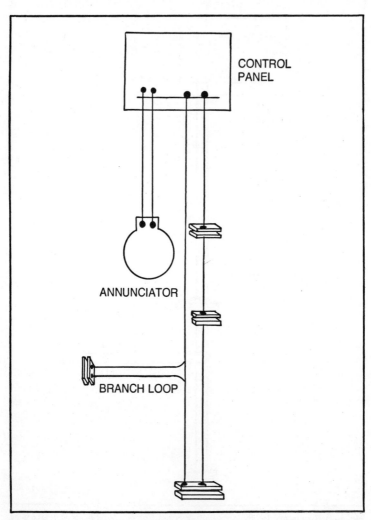

Fig. 14-2. The closed circuit alarm system consisting of normally open detectors wired in series.

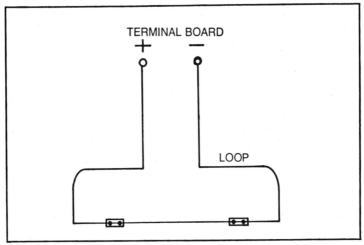

Fig. 14-3. Closed circuit system wiring.

both terminals of the detectors. The other wire remains intact. Polarity will be observed, especially when called for on double circuit detectors such as floor mats, foil glass or window screens. In any of these, connect both conductors to the two wire leads of the detector, then connect the new end of the circuit wire to the two exit leads of the detector. But here polarity must be closely observed. The positive line going in is reconnected to the positive line of the continuing circuit wire (Fig. 14-2). Then continue the circuit wires to the next detector, resuming the original format of one wire connected to both terminals in series and the other wire left intact. At the end of the line, bring the circuit wiring loop to the control panel and wire to terminal board (Fig. 14-3).

INSTALLING BRANCH LINES

At times it becomes necessary to run wire to a detector where it would be difficult to run the circuit loop. In that case, simply tap into the loop on one side (Fig. 14-4) and run a branch line to the detector. Always ensure that the branch line is secured to the main loop. To ensure a permanent connection, either use wire nuts or solder it.

SHUNT SWITCH

The *shunt switch* (key or switch operated) allows the security alarm to remain activated when entering or exiting through the shunted entrance. The switch does not turn off the system, but only

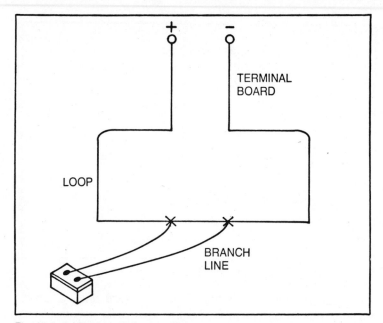

Fig. 14-4. Adding branch line to wiring.

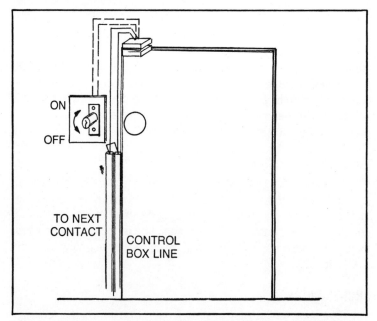

Fig. 14-5. The shunt switch allows the security alarm to remain activated when entering or exiting through the shunted entrance.

controls the detector at that particular door. The shunt switch sets the system as required (open or closed, depending on the type of system used) when using that entrance (Fig. 14-5).

Glossary

ace lock—A locking mechanism using a circular key and in which the tumblers are arranged in a circle.

active door—In a pair of doors, the active door is the one which must be opened first.

antipick latch bolt—A small parallel bolt which is depressed by the strike when the door is closed; it automatically deadlocks the spring latch bolt, preventing it from responding to external pressure such as a shove-knife or ice pick.

armored front—A steel plate applied to the regular front of a lock by machine screws; the plate protects the cylinder set screws so they cannot be loosened without removing the armored front.

back plate—Affixed to the inside of a door. It has holes which allow screws to enter the rear of a rim lock, holding it in place.

backset—The horizontal distance from the door edge to the centerline of the cylinder; or, if so stated, to the keyhole or thumb knob.

barrel key—A round stem key, hollow at the end. The hollow end fits over a post affixed to the interior lock casing. Sometimes called a hollow post key.

bevel of door—The angle of the lock edge of the door in relation to the inside and outside surfaces of the stile. While thin light doors are not usually beveled, heavier doors are normally beveled ⅛ inch in two inches.

bevel of lock—Used to designate the direction of the bevel of the latch bolt. The bevel of the lock is determined by the direction in which the door opens. If the door opens away from the key-operated side, the lock is a regular bevel. If the door opens toward the key operated side, the lock is a *reverse bevel*.

bit—The blade projecting from the stem and containing the cuts or notches which contact the tumblers or bolt operating the lock. That part of the key inserted first into the keyhole.

bitting—Arrangement of a series of cuts in the bit, made in a manner to move the tumblers to the release position.

blank—A key blank that is cut as needed.

bolt—The part of the lock which slides out of the lock case and fits into a recess or strike plate in the door frame, locking the door.

bow—The part grasped by the hand or fingers when the key is in use.

box strike—A strike which provides complete housing for the bolts, for both neater application and greater protection.

burglar proof—A lock or safe so designed as to be absolutely impregnable by a thief without the use of explosives, special tools and unlimited time. Usually approved and certified by Underwriters' Laboratories and affixed with a label.

burglar-resistant—A lock or safe designed to be capable of resisting attack by thieves for an allowable period of time. It may eventually be defeated.

butt hinge—A hinge designed for application to the butt edge of the door.

cam—A rotating piece attached to the end of the cylinder plug to operate the locking mechanism or bolt.

case—The box or container housing the locking mechanism.

casement window—One with sash hinged at the side.

case ward—An obstruction molded or cast as part of the lock casing of a warded lock.

change key—A key which will operate only one lock in a series, as distinguished from a *master key* which will operate all locks in a series.

changes—The various tumbler arrangements a particular lock model will accommodate and still differentiate between keys.

clean opening—A skilled entry; to open a key or combination lock without forceful entry.

clevis—A metal link used to attach a chain to a padlock.

code—A record of a key bitting arrangement corresponding to a particular key change number.

combination lock—A keyless locking mechanism requiring that certain internal parts be placed in correct alignment through the use of an outer numeral or letter dial which, when rotated to the authorized digits or letters, opens the lock.

communicating door lock—A lock having a latch bolt operated by knobs and a dead bolt operated by a turn knob.

connecting bar—A flat or triangle shaped spindle attached to the rotating plug of a rim lock mechanism to operate the bolt.

core—The round rotating portion of a lock cylinder, containing the keyway (also called the *plug*).

curved lip strike—A strike with its lip curved to conform to the detail of the door casing.

cuts—A term referring to the indentations made in the blade of a key blank in order that the key may operate the lock (sometimes called *bitting* or *notches*).

cylinder—The housing (sometimes called the shell) containing the plug, pin tumblers, springs and keyway.

cylinder collar—A plate used under the lip of the cylinder, often referred to as *cylinder rings*.

cylinder lock—A lock mechanism operated by a cylinder instead of levers, wards or other types of lock designs.

cylinder screw—The set screw in the front of a cylinder lock which prevents the cylinder from being turned after installation.

deadbolt—A lock bolt having no spring action, usually rectangular in shape and actuated by key or turn knob; it becomes locked against end pressure when projected.

dead latch—A lock with a beveled latch type bolt which can be automatically or manually locked against end pressure when projected.

dead lock—One having a dead bolt only.

depth key—A special key (enabling a locksmith to cut blank keys made for a special lock according to code arrangements.

disc tumbler—An oval shaped disc with a rectangular hole of differing depths corresponding to cuts in a key. A number of ·these are used in the core of a disc tumble lock (sometimes called wafers).

double bitted key—A key having bitting or cuts on two sides to actuate the tumblers of the lock.

double throw bolt—A bolt which, by automatic or manual operation, can be projected beyond its normal position, providing added security.

drill pin—A round pin or post projecting from the inside of the lock case opposite the key hole to receive a barrel or hollow post key (also called a post).

drivers—The upper set of pins in a pin tumbler cylinder which, activated by the springs, project downward into the plug until raised by insertion of the key to the shear line.

dummy cylinder—One without an operating mechanism, for use where effect is needed and to cover a hole previously drilled.

Dutch door— Consisting of two separate sections hung one over the other. Usually equipped so that both sections can be operated as one unit, or so that the upper section can be opened independently for ventilation.

easy spring—A term used in referring to the construction of a knob operated lock or latch.

electric release—Often called an *electric strike;* an electric strike plate which releases the bolt and the door when energized by pressing a button, using magnetic cards or special keys.

end ward—A ward on the inner case of a warded lock that requires end bit cuts on the key.

escutcheon—A plate, either protective or ornamental, containing openings for any or all of the controlling parts of the lock such as knob, handle, cylinder, keyhole and so forth.

face of a lock—The plate surface showing an edge of door after installation.

fence—Projecting piece, usually attached to the tail of the deadbolt, which passes through the gates of the lever tumblers when they are properly aligned, allowing the bolt to be retracted.

flat key—A thin flat stamped key, usually steel, and usually having square cut bitting on one or both sides.

following tool—The tool used to hold the upper pins and springs in place when removing the plug from the cylinder and replacing the plug (sometimes called a *universal follow-through*).

French door—A full length, glass paned door.

front of lock—The plate through which the ends of the bolts project.

front door lock—A lock for use on primary or entrance doors; usually a dead bolt and latch bolt with stop works for controlling operation of the latch bolt from the outside.

gates or gating—The opening in lever tumblers through which the fence passes to permit actuation of the bolt.

grand master key—A master key having access to many more doors than a master key.

graphite—A lubricant useful for the intricate parts of a locking mechanism.

hand of a lock—A term used to indicate the adaptability of a piece of hardware to right or left hand doors.

hasp—A fastening device (swinging metal band) consisting of a loop and a slotted hinged plate, usually holding a padlock.

hollow post key—A barrel key.

horizontal lock—One with its major dimension horizontal.

hub—The part of a lock through which a spindle passes to actuate the mechanism.

indicator—A device, usually an inward/outward moving button, used in connection with hotel locks to indicate whether room is occupied.

jamb—The inside vertical face of a door or window frame.

keeper—Same as a *strike*.

keyhole plug—A small pin tumbler lock mechanism designed to block the keyhole of a bit key lock.

key extractor—A device constructed from fret or coping saw blade or fishing hook, used to remove key blades broken off in the keyway.

key plate—A small plate or escutcheon having a keyhole only.

keyway—The opening in a lock cylinder into which the key is inserted.

knob—A projecting grip, normally spherical in shape, for operating a latch bolt (also called a *thumb knob*).

knob latch—A door latch having a spring bolt only and operated by knobs.

latch—A door fastening device (commonly with a sliding or spring bolt), usually having no locking function.

latch bolt—A beveled spring bolt, usually operated by knob, lever handle or thumb piece.

layout board—A length of board with a combination of grooves used to hold pin tumblers in order while the locksmith works on the lock.

lever tumbler—Flat tumblers having a pivotal motion actuated by the bitting of the key and controlling of the locking function.

lip of a strike—The projecting part on which the latch bolt rides.

locking dog—The part of a padlock which engages the shackle and holds it in the locked position.

lock set—Lock, complete with trim, such as knobs, handles, escutcheons, and so on.

lock stile—The stile to which the lock is applied, as distinguished from the *hinged stile*.

master key—A key with bitting arranged to operate two or more locks of different changes in a group, each lock being also operated by its own individual key.

master keyed lock—A lock so keyed that it can be furnished in a group, each lock being operated by its own individual or change key, and all locks in the group being operated by a master key.

master keying—The process of setting up the key changes in a group or a series of locks in a master keyed system.

master pin—A small pin or wafer pin used when complex master keying is necessary. A master pin produces two positions at which the lower pins reach the circumference of the plug, allowing the plug to rotate.

mortise—An opening made to receive a lock or other hardware.

mushroom pins—A mushroom shaped driver pin (also called a *spool pin*).

night latch—An auxiliary lock having a spring latch bolt and functioning independently of, and providing additional security to, the regular lock on the door.

nose plate—A small plate surrounding the face of the cylinder to give a finished appearance on certain types of cylinder locks.

paracentric—A term used to distinguish a milled cylinder key from other types, such as bit key, flat steel key, tubular key and so forth.

pick—A tool or device, other than the correct key, used to open a lock.

pin tumblers—Circular pins inside chambers within the lock cylinder. They work against coil springs and prevent the cylinder plug from rotating until the pins are raised to the exact height by the authorized key.

plug—The round part containing the keyway and lower pins. Also called the *core*.

plug retainer—The part which retains the plug in a cylinder lock.

rail—The horizontal members of a door which join the stiles. Normally top and bottom rails and the lock (center) rail.

reverse bevel—A term used to indicate that the bevel of the latch bolt is toward the inside of the door; opposite of regular.

reversible lock—Usually one in which the latch bolt can be turned over so as to adapt it to doors of either hand, opening in or out.

rim—A term indicating locks and other hardware designed for application to the surface of the door.

rounded front—A lock front conforming to the rounded edge of a double acting door. Usually of 2¼ inch radius.

security—The ability of a locking device to withstand attempts at unlawful entry.

set screw—One which, by checking another screw or other movable part, prevents it from loosening.

shank—The part of a key between the bow or handle and the bit or wing (also called the *stem*).

shear line—The circumference of the plug. That space between the shell and the plug where the driver pins and the lower pins meet.

shell—The circular outer section of a pin tumbler lock, housing the plug, lower and upper pins and springs (also called the *cylinder*).

snapper pick—A vibration pick constructed of spring steel and shaped like a safety pin. Used to pick pin tumbler locks.

spindle—The bar connecting the knobs or knob and passing through the hub of the lock for the purpose of transmitting the knob action to the latch.

stem—The round portion of the key extending from the bow and to which the bit is attached.

stop—The button of the lock which serves to deadlock the latch bolt against outside knob or thumb piece, or holds the bolt retracted, as in the case of a night latch.

strike or strike plate—The part of the lock recessed to receive the bolts of locks, latches or fasteners, and applied to the door frame. Also called a *keeper*.

stud—The rough frame that forms the "skeleton" of the inside of the house.

talon—The notch in the tail piece of a deadbolt into which the rotating member engages to throw or retract the bolt.

tension wrench—Tool inserted into the keyway and used to apply rotational (turning) pressure when picking a lock.

thumb knob—The small knob above the door knob intended to be pressed by the thumb to operate the latch bolt.

tubular lock—One having a cylindrically shaped case and requiring a bored hole rather than a chiseled rectangular mortise.

universal—A term used to describe a lock or other device which can be used on doors of either hand, without change.

ward—An obstruction projecting from the lock case or side of the key hole, intended to block entrance or rotation of an improperly cut key.

warded key—One having notches designed to clear the warding of the lock.

Index

Edited by Roland Phelps